CONSTRUCTION

METHODS FOR CIVIL ENGINEERING

SECOND EDITION

Errol van Amsterdam

Construction Methods for Civil Engineering

First published 2000
Reprinted 2003
Reprinted 2005
Reprinted 2008
Reprinted 2009
Reprinted 2011
Reprinted 2012
Second edition 2014
Reprinted 2015 (twice)

Juta and Company Ltd
First Floor
Sunclare Building
21 Dreyer Street
Claremont
7708

PO Box 14373, Lansdowne, 7779, Cape Town, South Africa

ISBN 978-0-70219-770-3

Project manager: Debbie Henry
Editor: Jenny de Wet
Proofreader: Lee-Ann Ashcroft
Typesetter: Trace Digital Services
Cover designer: Marius Roux
Indexer: Jenny de Wet
Printed in South Africa by Bevan Litho, Somerset West

Typeset in 10.5pt on 13pt Photina

Contents

A note to the student

This book does not aim to be a comprehensive reference book on all the methods followed in civil engineering construction, but is rather an introduction to the subject for students. It covers the curricula normally followed by universities of technology around South Africa. It provides a set of course notes with ample illustrations, questions and evaluations covering the basic knowledge you should acquire in the first year of the Civil Engineering Diploma course.

Civil Engineering is a career based on a wide variety of specialist fields, making it impossible to deal with any specific topic in great depth. The object of this course is to provide you, the student, with a sound theoretical background to enable you to apply this knowledge when observing work in progress and while you gain your own practical experience. Some of the topics in this book will be covered in more detail later in your Civil Engineering studies. We have also added more information in certain chapters, simplified others by providing worked examples and extended the section on labour intensive construction activities.

The language in the text is simple, conversational English. Difficult concepts and technical terminology are explained throughout the book.

The text is set out in such a way that you should be able to work through the book by yourself. New concepts are explained and reinforced by providing examples with solutions to work through. The figures throughout the text help you to understand the material and clarify concepts. Structured self-evaluation exercises appear throughout each chapter. The summary in each chapter enables you to see at a glance what you should have learnt in the chapter.

There are also activities for you to work through, which will allow you to make sure that you have understood the work you have just covered.

Check your answers to the self-evaluation questions by referring to those given at the end of each chapter *after* you have tried to complete them. Unless you do all the activities and work through the self-evaluation questions, you will never begin to master the content.

Icons page

We have used six icons in this book:

This is an activity icon. When you see this icon you will know that it is time to DO something!

Active and enjoyable, and these help you to understand the subject. Feel free to do them with a friend or group of friends.

The take note icon appears alongside all the extremely important information.

This is a definition icon. Read the definitions carefully because the details are important.

This is a self-evaluation icon. The self-evaluation questions enable you to assess your understanding of concepts discussed in that chapter. The solutions to these questions are given at the end of each chapter.

This is a reminder icon. It directs you to a previous section of work or chapter where the information was explained or dealt with.

This is a question icon. It examines a portion of the work you have completed by posing a direct question.

Acknowledgements

The authors and publisher wish to thank the following persons and institutions for their invaluable contribution to the development of this publication and for permission to reproduce material:

- The United States Agency for International Development (USAID), for funding the project. This Materials Development Project formed part of the Tertiary Education Linkages Project (TELP) which focused on capacity building at historically disadvantaged technickons through the establishment of linkages with universities in the United States of America.
- Contributors and moderators from the following South African institutions formerly known as: Mangosuthu Technikon, ML Sultan Technikon, Peninsula Technikon, Technikon Eastern Cape, Technikon Northern Gauteng, Technikon Southern Africa.
- Contributors and moderators from the United States University Consortium comprising Howard University, Massachusetts Institute of Technology, Clark Atlanta University, North Carolina A & T State University.
- The Council for Scientific and Industrial Research, South Africa (figs 3.4, 3.5); The Department of Forestry and Water Affairs (figs 4.1, 4.11, 4.12, 4.13); INPRA/Topham Pictures (fig 9.2); Mr LV Leech (fig 2.24); The South African Bureau of Standards (fig 2.3); The Southern African Institute of Steel Construction (figs 2.17, 2.21, 2.23).

Chapter 1 Earthworks

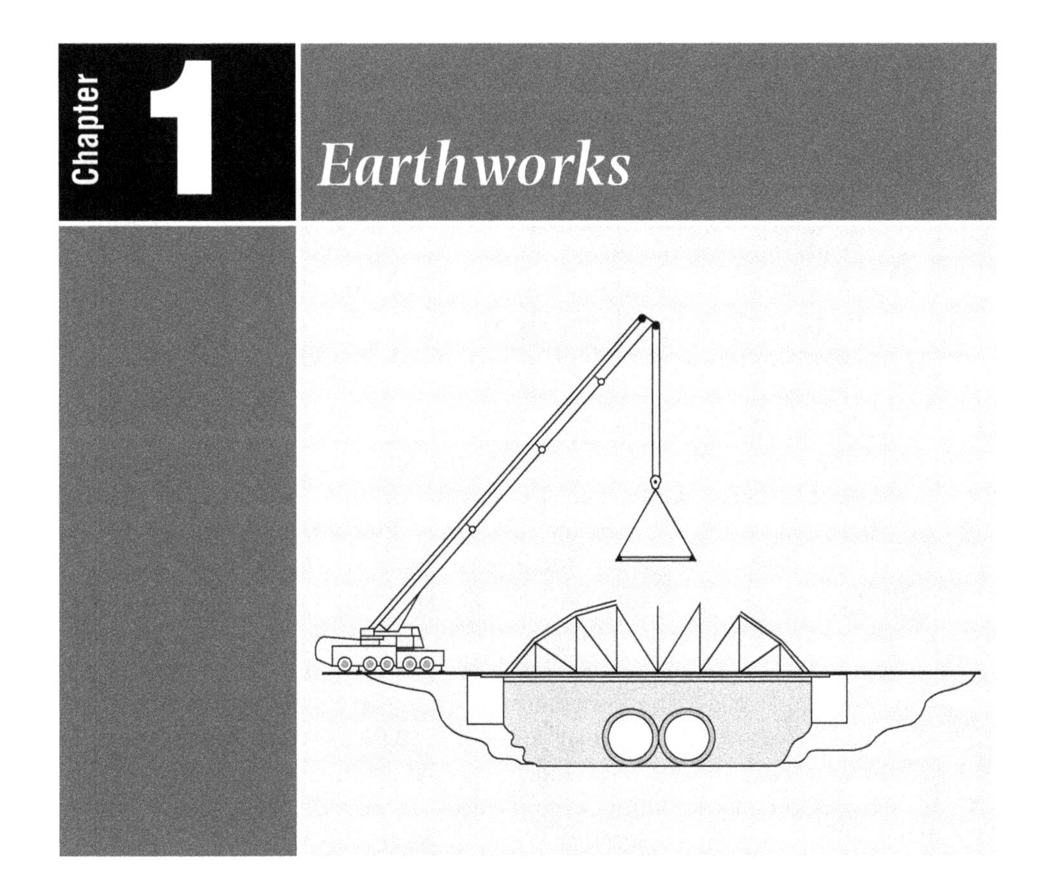

Outcomes

After studying this unit, you should be able to:

- Expand on your knowledge of soil properties as it applies to foundations
- Understand the difference between foundation types
- Calculate basic loads and determine minimum widths of foundations
- Explain excavations and methods of making it safe from collapse
- Identify how timbering is used to support the sides of an excavation
- Know some of the regulations pertaining to excavations
- Understand the various types of retaining walls
- Identify and explain the types of construction plant used on site as it pertains to earthworks.

1.1 Introduction

Earthworks in modern engineering construction can be interpreted in a variety of ways depending on where one is involved in the process. For example:

- a geotechnical (soils) engineer will interpret earthworks as dealing with soils in general;
- a structural engineer and building contractor will interpret earthworks with reference to foundations; and
- a roads engineer will look at general earthworks in relation to road layerworks.

Before any earthworks can start, the site and its boundaries must be established. Usually, the site and its location are identified, then a surveyor places the boundary pegs. Once the boundary pegs are placed, the site needs to be cleared and leveled where necessary. Imagine an area large enough to accommodate a major shopping mall – usually this can be anything between 20 000 to 30 000 square metres – and that is only counting the shopping area. Space must still be made for parking, roads, delivery of goods, accommodation of engineering services, etc. Go down to a large shopping area near you or otherwise use Google maps to define and establish this space. What you see is the final product, but consider that before any buildings were erected, the area on which the shopping mall is located had to be cleared and shaped to the structure you now see.

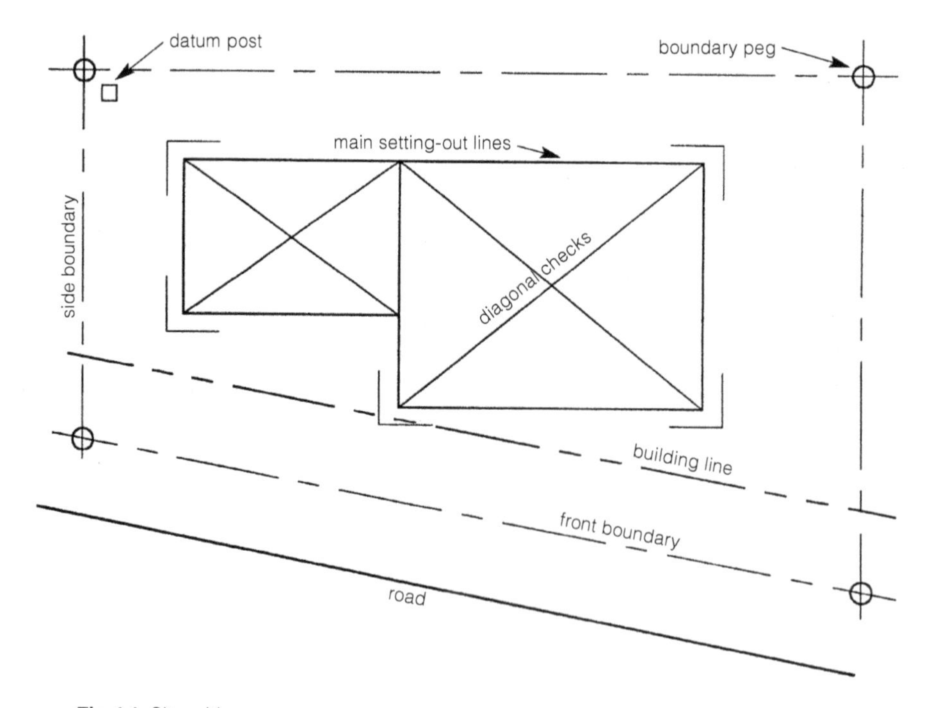

Fig 1.1 Site with pegs

1.1.1 Site clearance

The site has to be cleared of vegetation, bushes, trees, topsoil, disused buildings and any other unwanted features. The removal of trees and bushes can either be done manually or mechanically. **Construction plant** like bulldozers and front-end loaders are used for this purpose and all unusable material is removed from the site by truck.

Why do you think it is necessary to remove vegetation and other debris from the site? It is civil engineering practice to remove the topsoil layer (normally approximately 150 mm deep) from a site and either stockpile it for later re-use or remove this material completely off site. If you refer to the Construction Materials subject, you will understand that building on top of a material rich in vegetation and other growth makes for a very unstable 'foundation', hence it is better to remove such material.

Let's do a very quick calculation to give you an idea of **volumes of material**. Remember the shopping centre we discussed earlier. Assuming that the total area for the shopping mall is 60 000 square metres (taking into account all the space requirements), and if we need to remove only 150 mm of topsoil (an accepted norm in civil engineering), what is the total volume of material to be disposed of off site or stockpiled?

If you calculated 9 000 cubic metres of material, you are correct. How did we get this? Well, convert 150 mm to metres by dividing by 1 000 and then multiply the answer by 60 000 m^2.

If we consider a normal disposal truck can manage an average of 6 m^3 per load, how many trucks would be required to remove this volume of material? If you answered 1500 trucks, you are correct. Obviously, you do get trucks of higher loading capacity which will reduce the number of loads required, but you get the basic idea.

Construction plant is the term used to describe all machinery used on a construction site, ranging from excavators and compactors to haulers.

Activity 1.1

Arrange a trip to your nearest major construction site and take note of or otherwise take photographs of all construction plant utilised on site. Remember the size of the equipment used is not an indicator of whether it is classified as 'construction plant'. Once you have collated this information, use resources at your disposal, for example the internet, to help in identifying these plant. See Activity 1.8 for the outcomes associated with this task.

On site, bulldozers are often seen moving and stacking soil in a heap. These heaps are referred to as stockpiles and are located away from the normal construction operation so that they do not interfere with the daily routine.

The removal of existing buildings is best left to demolition experts. The removal of large trees can be just as dangerous and it is best to call in experts from the municipal or state forestry department.

It is spectacular to see a building being demolished using explosives. But do not try this on your own! It is dangerous and you could damage property and injure people. Leave it to the experts.

1.1.2 Setting out the site

Once the site has been cleared of vegetation and rubble, work boundaries, which are often different from the site boundaries, can be set. A site surveyor or technician sets out the work, be it a building, a dam or a road. Where necessary, a **datum level** must be established on the site so that all construction work can be related to this fixed point. For large works that stretch over vast areas – for example, roads and dams – several coordinated benchmarks will be established around the site, all referenced from the initial datum peg.

Why is it important to establish boundaries and levels for the project site?

Imagine playing sport without a set of rules in place or playing soccer without the boundaries of the field being established! The same applies to a construction site so that construction can take place within those confines.

It is important that the datum level is firmly grounded to maintain the fixed level, so the datum peg is usually set in concrete.

In this chapter we will examine some of the methods and plant used in excavation and earthworks in the civil engineering construction sector.

1.2 Definitions

- **Bearing capacity** is the safe load per unit area that the ground can carry, measured in kN/m^2.
- **Bearing pressure** is the pressure produced on the grounds by the applied loads in kN/m^2.

- **Settlement** refers to ground movement and can be caused by deformation of the soil due to the loads imposed on it.
- **Backfill** is material from the site that, if suitable, will be used to fill in around the walls, foundations and any pipe trenches.
- **Self-weight loads** are all the permanent components of a structure – for example, foundations, walls, columns, beams and slabs.
- **Imposed loads** are non-permanent components of a structure – for example, people, furniture and vehicles.
- **Differential settlement** is the uneven settlement of a structure due to load distribution or poor soil conditions.

1.3 Foundations

Foundations are built below the ground and form the base for any structure. Their function is to distribute the load of the top structure into the soil. Foundations are also referred to as bases or footings. Have you passed a construction site where houses are being built? You may have noticed trenches being dug in the earth and then filled with concrete. Once the concrete has set, the bricklayers can start laying the bricks of the visible structure.

A similar method is used for multi-storey structures and bridges, except that the foundations vary in shape and size. There are two categories of foundations: **shallow foundations** and **deep foundations.**

Shallow foundations transfer the loads of the structure to the soil at a point near to the ground floor of the building. Shallow foundations are usually classified as less than 1.5 m deep.

Do you remember how shallow the foundations were at the construction site you viewed?

Deep foundations transfer the load of the structure to the soil some distance below the ground floor of the building. Deep foundations are classified as being deeper than 3.0 m.

Activity 1.2

If shallow foundations are less than 1.5 m deep and deep foundations are deeper than 3.0 m, what happens between these depths? Consult professionals either on site or in a design office to find out the answer.

1.4 Foundation types

Foundations can be classified in many ways, but the most common reference is their form. There are five basic foundation types:

1. **Strip foundations** are used where light loads need support – for example, in supporting walls, in house construction. A reinforced concrete strip foundation can be used for heavier loads.
2. **Raft foundations** are used where the soil has a low bearing capacity or undergoes **differential settlement** (unequal) and where light to medium loads are supported. Raft foundations are also used for buildings with basements. Raft foundations generally have a larger surface area over which to spread the load.

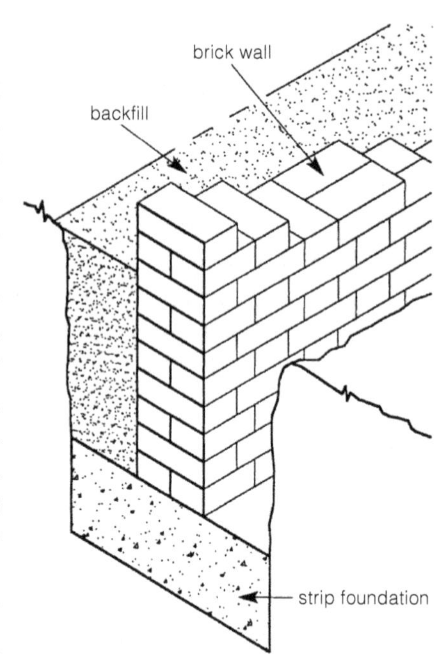

Fig 1.2 A typical strip foundation

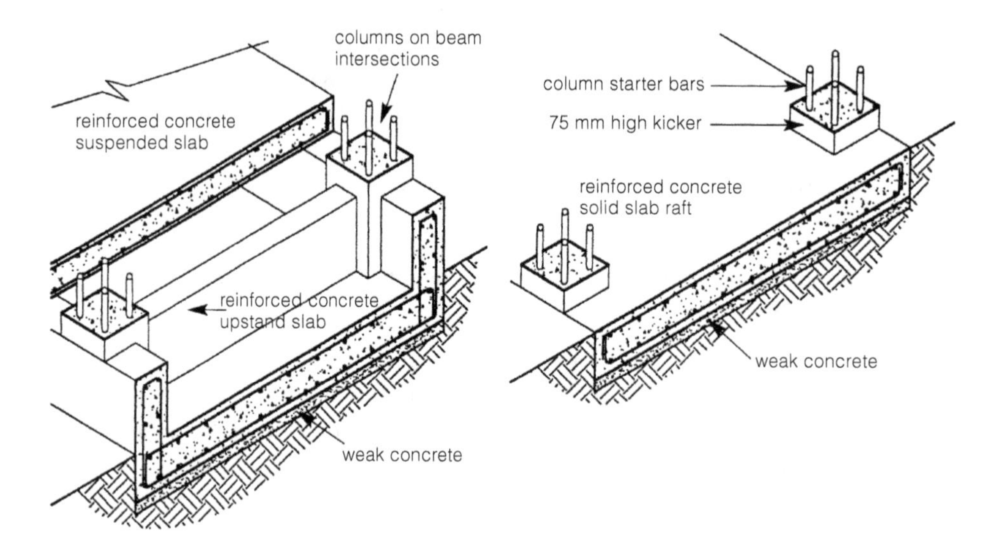

Fig 1.3 Beam and slab raft foundation

Fig 1.4 Solid slab raft foundation

3. **Pad or isolated foundations** are commonly used in portal frame construction and for columns in multi-storey structures and bridges.
4. **Pile foundations** are used for structures where the loads have to be transmitted to a point some distance below ground level. Pile foundations are often used where ground conditions are poor or

the bearing capacity is very low. Piles are driven down to a level where conditions are more suitable. Pile foundations generally are able to carry much higher loads then other types of foundations. Bridge foundations are often constructed using piles. A pile cap is formed, which then becomes part of the support, whether it is for a pier or an abutment.

5. **Caissons** are box-like structures that can be sunk through ground or in water to install foundations or similar structures below the water line or water table.

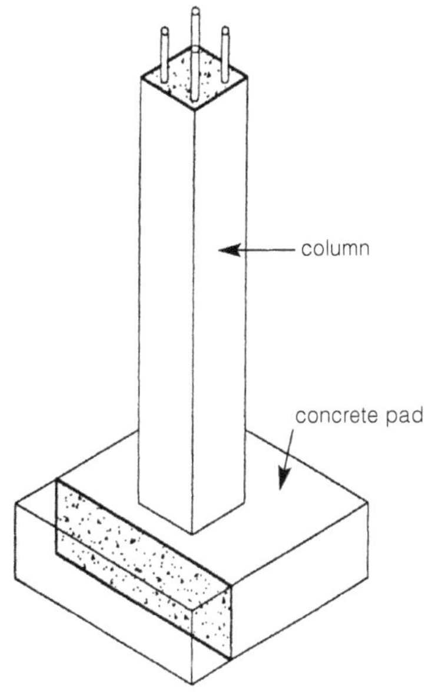

Fig 1.5 Typical pad foundation

The feature of using these foundation types is that they are constructed above ground or water level and then sunk as a single unit to the required depth. One can correctly assume that this assembled unit will eventually form part of the permanent works. Caissons are suitable for constructing foundations in water or through unstable shifting soil to depths greater than 25 m.

There are several caissons used in constructing foundations:

- box caissons;
- open caissons; and
- pneumatic caissons.

Box caissons are open at the top and closed at the bottom. They are designed to be sunk onto a prepared foundation below water level where erosion cannot undermine them. Box caissons are most suited for compacted inerodable gravel

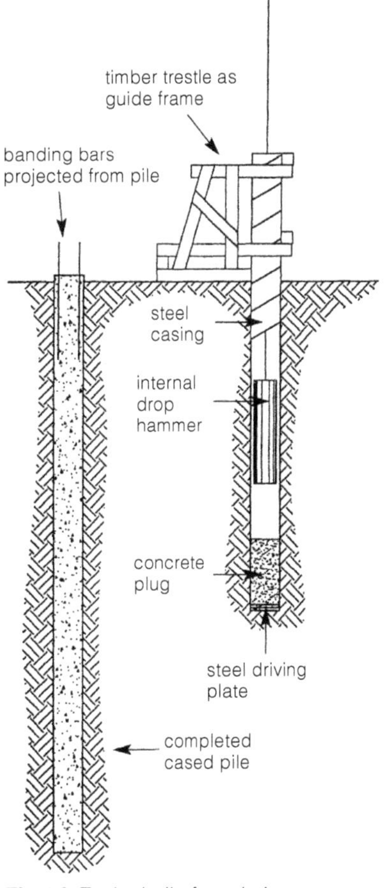

Fig 1.6 Typical pile foundation

or rock, but can also be used on uneven rock surfaces if all the loose material is removed and replaced by a blanket of sound crushed rock.

Why do you think that box caissons should not be constructed on loose erodible soil?

This is to prevent the caissons from settling and being unstable, which would not provide a good foundation.

Open caissons, as the name suggests, are open at the top and bottom, and sunk through excavation of the soil inside the caisson. Soil suitable for this method of construction includes soft clays, silts, sands and gravels. On reaching the required foundation level, open caissons are sealed by depositing a layer of concrete underwater in the bottom of the wells. The wells are then pumped dry and more concrete is placed, after which the caisson can be filled with clean sand or concrete.

Pneumatic caissons have a working chamber in which air is maintained above atmospheric pressure to prevent ingress of water into the excavation. Pneumatic caissons are used in preference to open well caissons in situations where excavations in unstable soil conditions will result in the possible settlement of adjacent structures. They are also used in variable ground or through ground with obstructions, where an open caisson would tilt or refuse further sinking. Excavations inside pneumatic caissons can take place under relatively dry conditions and, once the necessary level is reached, the sealing concrete is placed.

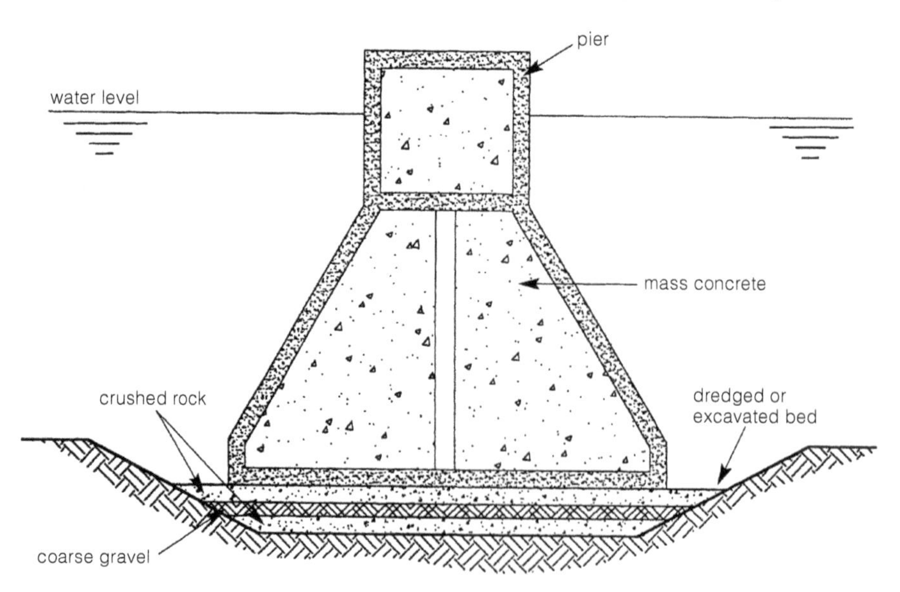

Fig 1.7 Box caisson

The table below indicates which of the five foundation types are deep or shallow.

Table 1.1 Foundation typology

Shallow foundations	Deep foundations
strip foundations	pile foundations
raft foundations	caissons
pad foundations	

1.5 Choosing foundation types

The choice and design of foundations for structures depends on the following factors:

- the total load of the building;
- the nature and bearing capacity of the soil; and
- the settlement of the soil.

The total load of a building is taken per metre run and calculated for the worst case scenario. The data required includes:

Self-weight loads Roof load and wind loads
Floor loads
Wall, column, slab and beam loads on the foundation
Imposed loads People, furniture, etc.

Total load = self-weight loads + imposed loads

Remember, 'self-weight load' is the term used to describe all the components of a building that, once placed, will remain in that position as long as the building exists. To establish the self-weight load, one needs to calculate the mass of the components.

Do you remember the chapter on concrete in the *Materials* handbook? Concrete has a density of 2 400 kg/m^3 and this figure must be used to calculate the self-weight load.

Activity 1.3

Calculate the self-weight load (in Newtons per metre run) for the concrete wall in the sketch below.

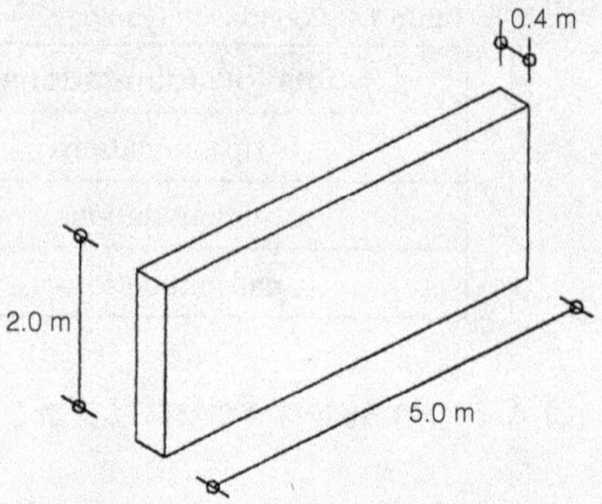

Solution

Density = mass/volume

Volume = L × B × H
= 5.0 m × 0.4 m × 2.0 m
= 4.0 m^3

Mass = density × volume
= 2 400 kg/m^3 × 4.0 m^3
= 9 600 kg

Weight mass × gravitational acceleration
= 9 600 kg × 9.81 m.s^2
= 94 176 kgm.s^2
= 94 176 N

Self-weight load per metre run = self-weight load (mass of concrete)/length of wall
= 94 176 N/5.0 m
= 18 835.20 N/m run

Fig 1.8 Dimensions of concrete wall

1.5.1 Determining the nature and bearing capacity of soil

The nature and bearing capacity of the soil can be determined by:

- trial holes and trial pits;
- boreholes and core analysis;
- local knowledge;
- soil testing methods like the California bearing ratio (CBR), nuclear testing methods, etc.;
- *in situ* field tests like plate loading tests, standard penetration tests, dynamic cone penetrometer (DCP), etc,; and
- laboratory testing – for example, the triaxial test.

The design of the foundation can be considered only after determining the nature of the soil, its bearing capacity and possible settlement, and the total load exerted by the structure. The width of strip foundations for houses can be worked out using the following calculation:

$$\text{Minimum width (m)} = \frac{\text{total load of building per metre run}}{\text{bearing capacity of the soil}}$$

Activity 1.4

Calculate the minimum width of the foundation required to support a load of 1 260 kN per metre run, if the bearing capacity of the soil is 1 800 kN/m^2.

Solution

Total load per metre	= 1 260 kN
Bearing capacity of soil	= 1 800 kN/m^2
Therefore the minimum width required	$= \frac{1260\text{ kN/m}}{1\ 800\text{kN/m}^2}$
	= 0.70m

So, the width of the foundation strip must be 700 mm.

Normally, one calculates the self-weight load of the building by taking into account all the components of the building – i.e. calculating the area per metre run of brickwork or concrete and multiplying this by their respective densities.

The density of concrete = 2 400 kg/m^3
The density of bricks = 1 700 kg/m^3
Included or added to the self-weight load is the imposed load, which usually varies between 1 and 2 kg/m^3.

The bearing capacity of the soil can be calculated by testing the soil at a geotechnical laboratory or with field tests.

Don't forget to multiply by 'g' to get the force.

1.6 Excavations

Before foundations can be laid, a trench of the required depth and width needs to be excavated. On small contracts this is still carried out by hand, but on large works it is often more economical to use some form of mechanical excavator.

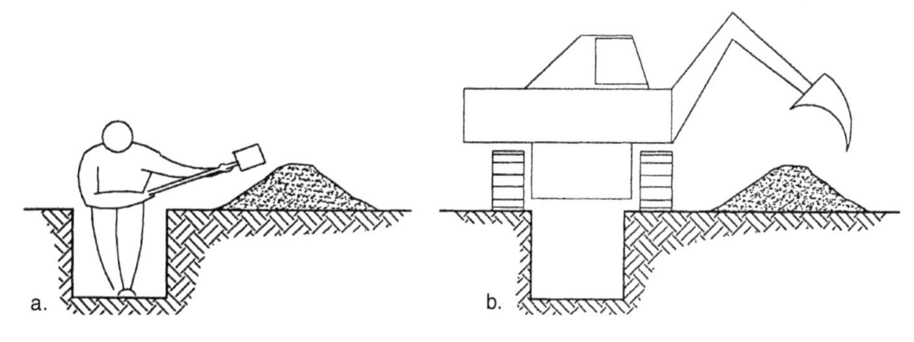

Fig 1.9 Digging a trench (a) by hand and (b) using an excavator

Doing excavation by hand lends itself to the application of labour-enhanced construction which is seen as a form of employment creation. You will read more about labour-enhanced construction methods later.

1.7 Timbering

Timbering refers to the temporary supports for the sides of excavations. It is also called planking, strutting, shuttering and shoring. The supports are used to:

- prevent the sides collapsing;
- prevent the inflow of loose material;
- prevent damage to the adjoining property;
- keep the excavation open by acting as a retaining wall to the sides of the trench; and, most importantly,
- keep the workers in the trench safe from collapsing soil.

Timbering is not needed for trenches less than 1.50 m deep if the soil is stable enough to support itself. However, if there is enough space around the proposed trench, the sides can be flattened out to prevent them from collapsing. It is common practice for the slope on either side of the trench to have a 1:4 gradient.

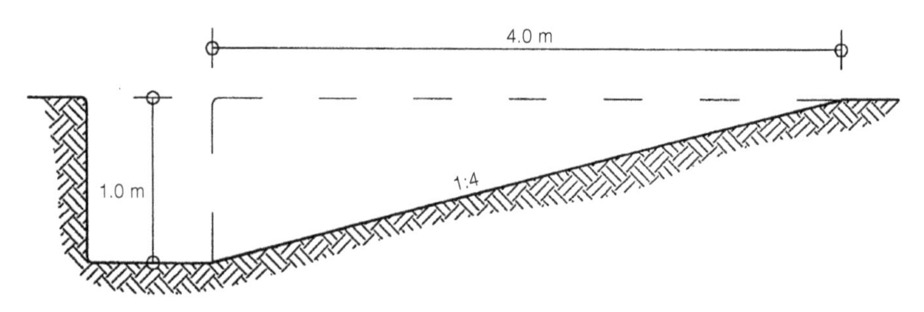

Fig 1.10 Unprotected trench

Activity 1.5

Use a ruler to mark out a slope of 1:4 in scale 1:50. Make a mark on an A4 page 20 cm along the horizontal axis and 5 cm along the vertical axis. Connect these two marks with a straight line. A similar exercise is done on site using timber to indicate the slope of the sides of the trench.

The contractor or a qualified engineer usually does the design of the timbering. The type and amount of timbering needed depends on:

- the nature of the soil (hard, firm, loose, sandy, clayey, etc);
- the depth of excavation;
- the presence of ground water;

- the proximity of buildings;
- weather conditions; and
- the duration of the operation.

1.8 Basic timbering excavation terminology

- **Poling boards** are 1.0 to 1.5 m in length and vary in cross-section from 175 × 38 mm to 225 × 50 mm. They are placed vertically and support the soil at the sides of the excavation.
- **Walings** are longitudinal members running the length of the trench and are used to support the poling boards.
- **Struts** are square timbers (100 × 100 mm or 150 × 150 mm) that are generally used to support the walings. They are placed approximately 2.0 m centre-to-centre to allow for adequate working space between them.
- **Sheeting** consists of horizontal boards abutting one another to provide a continuous barrier when excavating in loose soil.
- **Runners** are poling boards in continuous formation with tapered base edges. A common size is 225 × 50 mm. They are suitable for use in loose or waterlogged soils.
- **Puncheons** are vertical supports that are wedged between the walings at or near the ends of struts. They are necessary in deep excavations to prevent the walings from dropping/falling off.

1.9 Trench excavation safety

1.9.1 Causes of collapses

There are many reasons why trenches collapse, some of which are listed below.

- **Materials too close to the sides.** When working in a trench, the materials required (concrete, pipes, stones, etc.) are practically on the edge of the trench. This could result in the trench caving in if there are no supports.

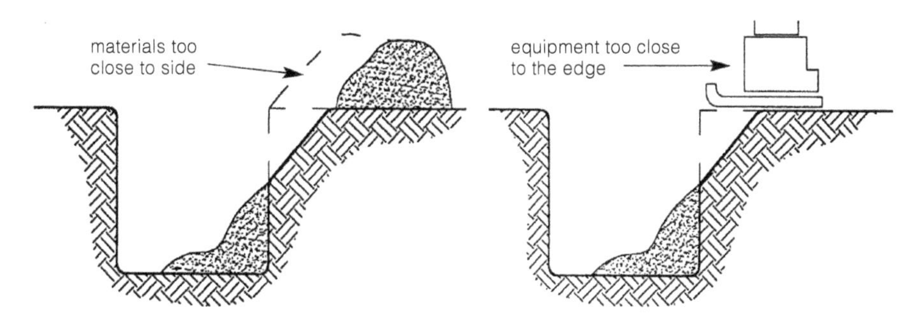

Fig 1.11 Causes of collapses

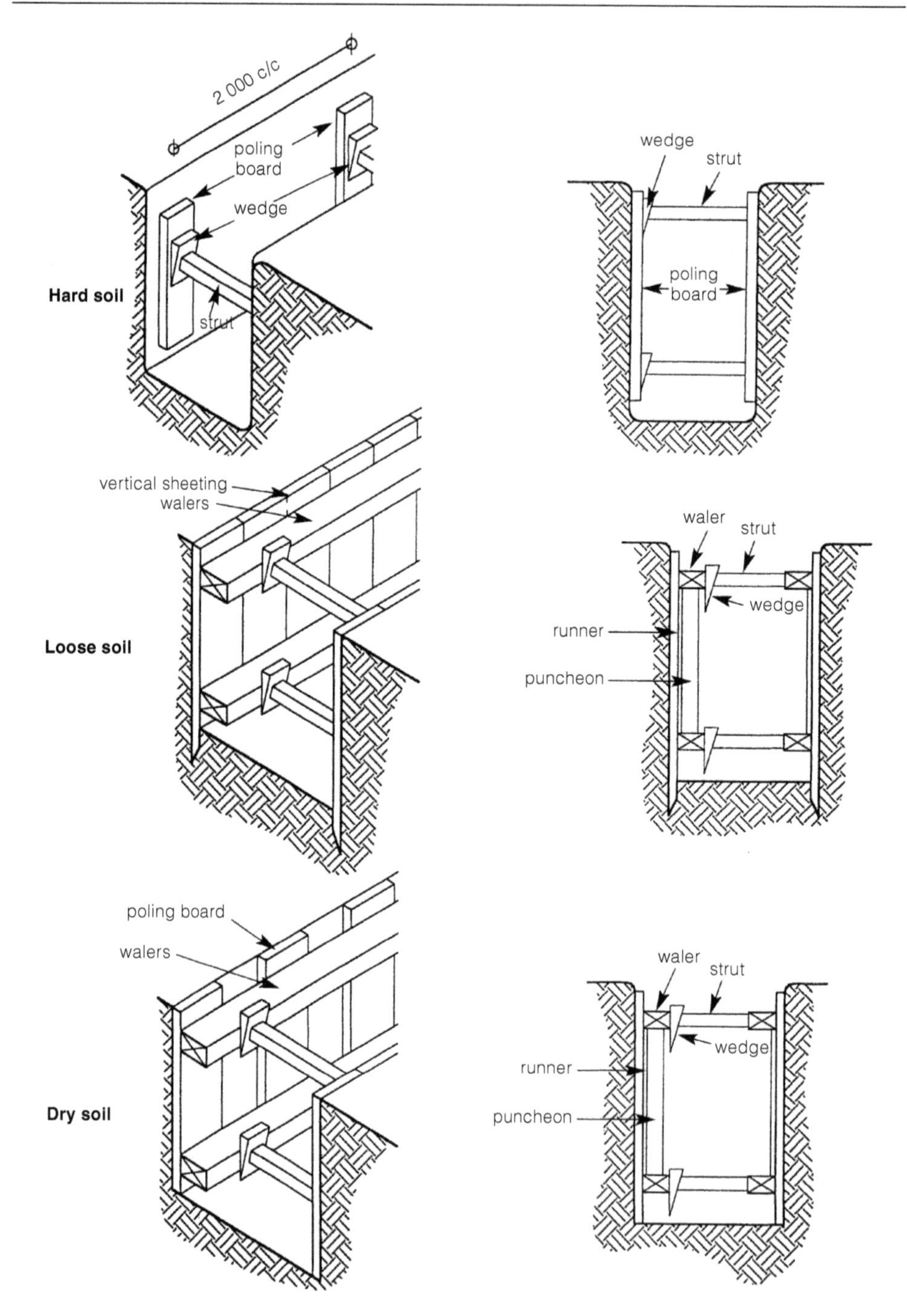

Fig 1.12 Timbering in hard, loose and dry soils

- **Excavation plant and equipment too close to the edge.** Operators of excavation plant often get too close to the edge of the trench. Also, construction workers tend to place compaction equipment, like plate compactors, near the edge of the trench, making it unstable.

- **Unstable subsoil material** (subsoil refers to the soil below the topsoil).
- **Variations in the nature of the soil**, for example pockets of sand.
- **Changes in the moisture content of the soil.** Drying out or rain may result in a change or breakdown of the soil strength.
- **Vibrations from compaction plant or passing vehicles** in the vicinity of the excavated trench.
- **The soil is unable to support its own weight.**
- **Unstable nature of the soil.** When excavating on or near the site of a previous excavation, the unstable nature of the soil could also result in collapse.

Because of the dangerous situations that can arise from unprotected trenches, the government, through the Department of Labour, has laid down regulations to protect both the public and workers.

Trench excavation regulations

1. Where trenches are close to the public, the following precautionary measures must be taken:
   a. All trenches must be protected with a fence.
   b. Red warning lights must be placed at regular intervals to warn the public.
2. No workers are allowed in trenches deeper than 1.5 m without the sides being protected or braced.
3. All bracing must be supported by cross bracing.
4. Bracing must be strong enough to support the soil walls.
5. No materials other than those required for the execution of the work are allowed in the trench and they are only allowed when required.
6. Any trenches exceeding 1.5 m must be given safe access – for example, a ladder.
7. All underground services must be investigated and located before major excavation begins.

If you strike a water main, the trench will flood almost immediately. If you strike an electrical cable, it may result in electric shocks. Damaging a telephone cable will result in a communication system breakdown. All of these are safety hazards and will mean loss in time and production as well as expensive repair costs.

8. Regular inspection of the trenches must occur, usually at daily, weekly or monthly intervals, or after bad weather and rain.
9. A responsible and competent person must carry out the inspection – preferably whoever is in charge of the shift or the work to be carried out.

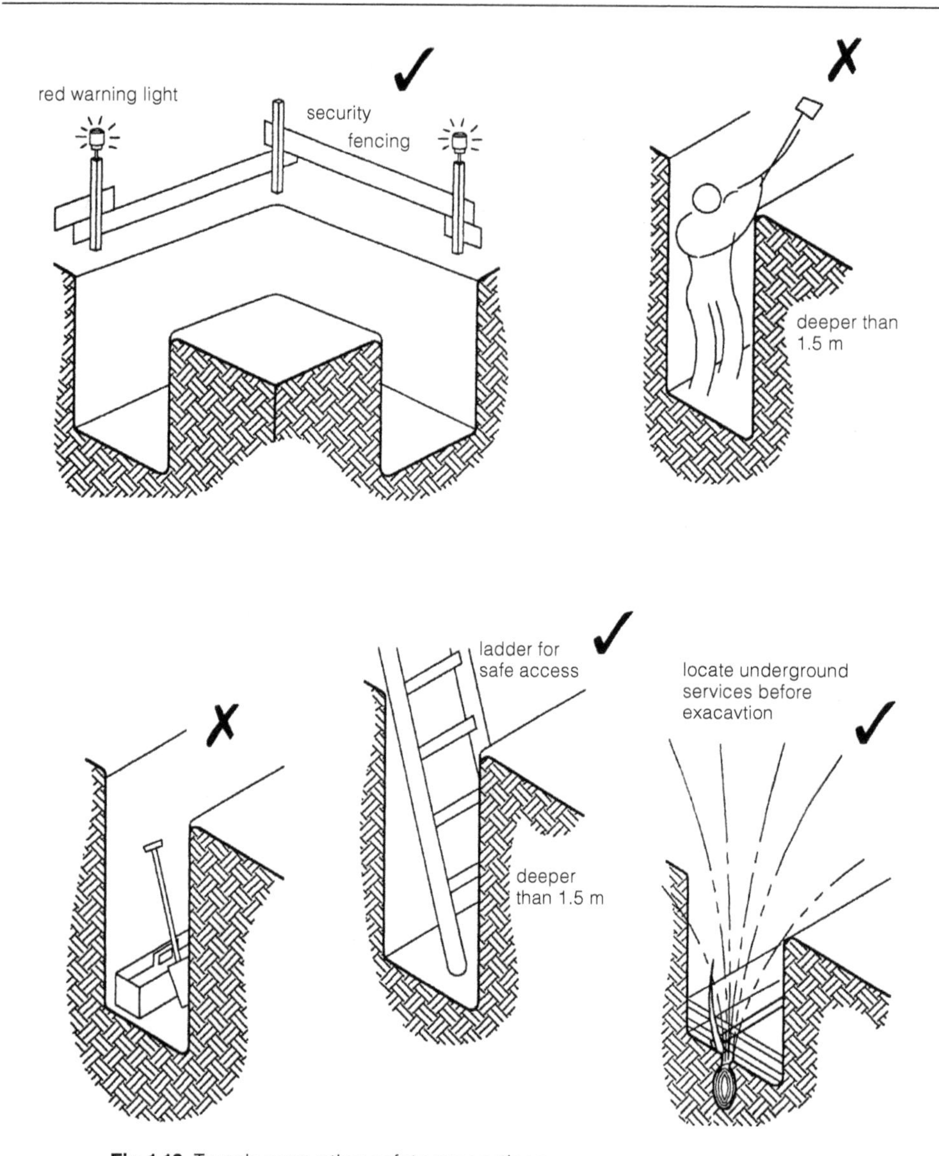

Fig 1.13 Trench excavation safety precautions

Activity 1.6

Safety in construction, particularly during excavation, is very important. Form a group with other students and access the library, newspapers and the internet (where possible) and see how much information is available on this topic. Use some of it to compile a report of approximately six pages depicting your findings as well as make a poster displaying a summary. When you next pass a construction site or make a site visit, see if you can spot which safety measures are being employed.

1.10 Excavating basements

Most multi-storey structures found in cities – for example, shopping malls and office blocks – require basements for various reasons including parking. Many of these can be very deep below the natural ground level.

Earthworks also need to be done for the construction of deep basements.

Four basic methods can be used:

- perimeter trench;
- raking struts;
- cofferdams, and
- diaphragm walls.

Do you remember that deep foundations are greater than 3.0 m below natural ground level?

1.10.1 Perimeter trench

This method is used where there is weak soil and a perimeter trench is excavated around the proposed basement excavation. The width and depth of the trench must be sufficient to accommodate the timbering, the basement retaining wall and adequate working space. Timbering as per the other methods described can be applied. To provide for a firm base, the base of the trench is covered with a 50–75 mm layer of weak concrete, commonly referred to as **blinder** or a **blinding layer**.

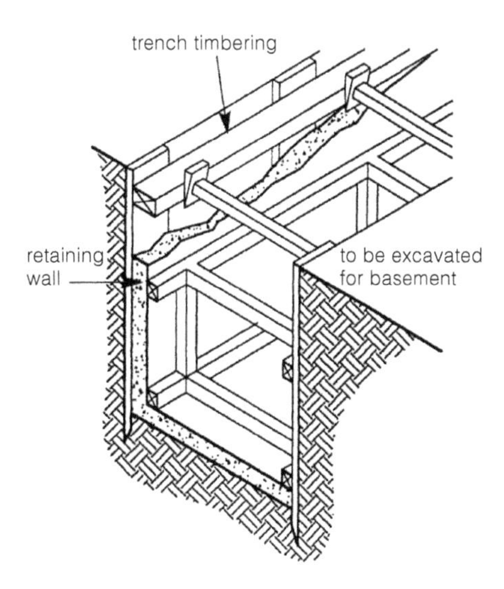

Fig 1.14 Perimeter trench excavation

1.10.2 Raking struts

This method is used where the basement area can be excavated in sections back to the perimeter line.

Firm subsoil must be present. The perimeter is trimmed and the timbering positioned and strutted using raking struts onto a common platform that is adequately braced.

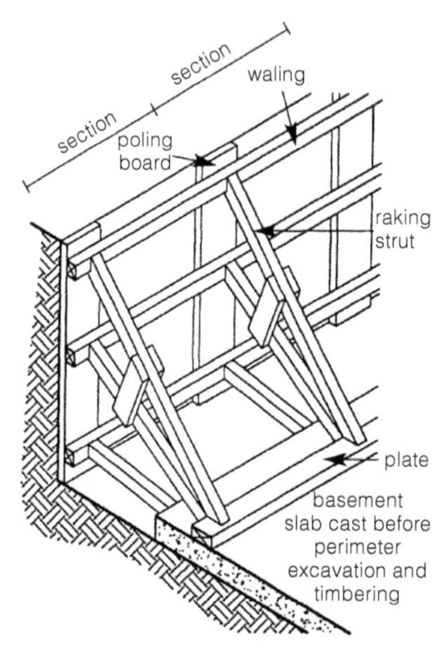

Fig 1.15 Raking struts

1.10.3 Cofferdams

The term 'cofferdam' comes from the French word *coffer*, meaning 'a box'. A cofferdam consists of a watertight enclosure, usually of interlocking steel sheet piles, used in waterlogged sites or in water. The cofferdam is constructed by placing the sheet piles in position, bracing them with tie rods and filling between the sheet piles. Excavation can then take place behind the cofferdam. Any water that enters the enclosure can be extracted using a pump.

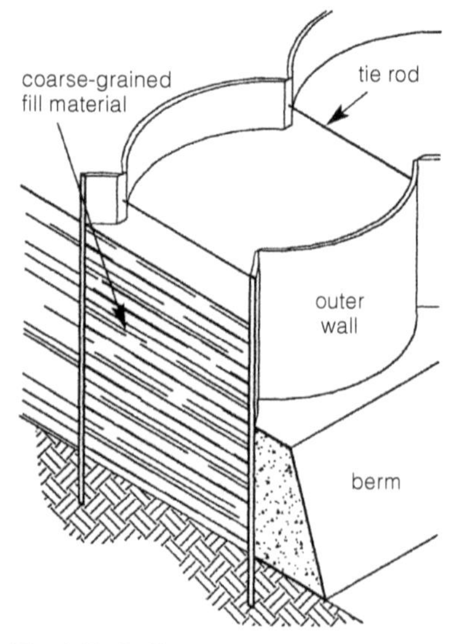

Fig 1.16 Cofferdam excavation

1.10.4 Diaphragm walls

A **diaphragm wall** is a type of dividing or retaining wall or structure that is used to retain large masses of soil.

The general method used to construct diaphragm walls is the bentonite slurry system. Bentonite is a form of clay that, when it comes into contact with water, forms a gel if left undisturbed. The basic procedure is as follows:

1. The excavated spoil material is replaced with bentonite slurry.
2. The slurry, when applied to the sides of the excavation, forms a soft gel that penetrates slightly into the soil.
3. This acts like a stopper, keeping the sides stable and groundwater out.
4. Panels to support the sides of the excavation can be inserted and excavation to the required depth can continue in this manner.
5. Once the required depth has been reached, a reinforcing cage is lowered into the hole and fixed in position.
6. A high slump concrete is poured into the trench and displaces the slurry which is either pumped away or recycled.

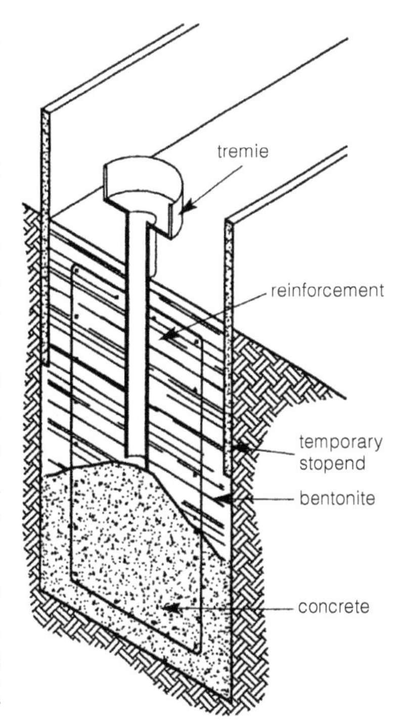

Fig 1.17 Diaphragm wall

1.11 Retaining walls

Normal soil has an angle (to the vertical) at which it can support itself without collapse – this angle is commonly referred to as the **'angle of repose'**. To assist the soil and make it stable, a retaining wall is constructed. The basic function of a retaining wall is to retain soil at an angle greater than it would naturally assume, usually at a vertical or near vertical position.

The natural slope, known as the angle of repose, of soil is measured in relation to the vertical and ranges from 37° for dry sand and gravel to almost 0° for wet clays.

The design of any retaining wall is basically concerned with the pressures of the retained soil and any subsoil water. Retaining walls must be designed to ensure that:

- overturning does not occur;
- sliding does not occur;
- the soil on which the wall rests is not overloaded; and that
- the materials used in construction are not overstressed.

The most difficult aspect of geotechnical engineering is to accurately define soil properties and their behaviours in certain circumstances due to the variability of material or conditions. When designing a retaining structure, the geotechnical engineer or technician must be able to calculate the pressure exerted by the soil at any point of the wall. To do this, several factors must be taken into account, including:

- the nature and type of soil;
- the height of the water table;
- subsoil water movements;
- the type of wall; and
- the materials used in the wall's construction.

Reinforced earth refers to the strengthening of soil used as fill material through the addition of strong tensile reinforcement in the form of strips. The strength is obtained through the generation of frictional forces between the soil and the reinforcement. The soil is compacted in layers, with reinforcing strips placed between the layers. Reinforced earth is being used successfully for ground slabs and foundations, dams and embankments, and walls.

The actual design and calculations for retaining structures will be presented in the Geotechnical Engineering and Structural Design courses in your later years of study.

There are various types of retaining walls, the most common being:

- gravity retaining walls;
- cantilever walls;
- precast concrete retaining structures; and
- concrete block retaining walls (CRBs), which are similar to precast concrete retaining structures.

1.11.1 Gravity retaining walls

These walls are also called 'mass retaining walls' and rely on their own mass and the friction on the underside of the wall to overcome the tendency to slide or overturn. They can be constructed from engineering bricks, natural stone or reinforced concrete.

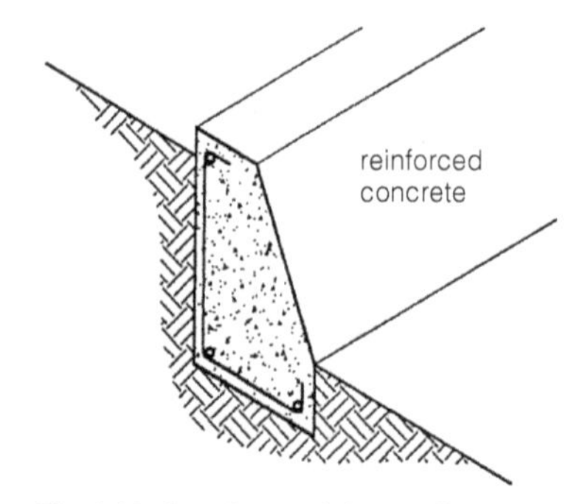

Fig 1.18 Gravity retaining wall

1.11.2 Cantilever walls

These walls are normally constructed of reinforced concrete. Two basic forms can be considered – a base with a large heel so that the mass of earth above can be added to the mass of the wall for design purposes, or, if this is not practicable, a cantilever wall with a large toe, or even both. Reinforced concrete retaining walls can be constructed using these forms.

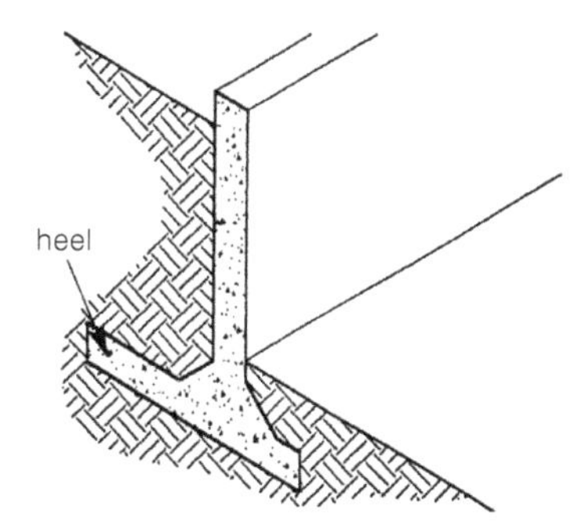

Fig 1.19 Cantilever retaining wall

1.11.3 Precast concrete retaining structures

There are many forms of precast concrete units available, but all adopt the same principle: that of keeping earth back. Precast concrete units of various shapes and sizes are available, but the most common are usually 600 mm wide and made from high grade concrete. They are erected on foundations and rely on their mass, interlocking ability and friction within the soil mass (due to metal strips attached to the individual units and tied back into the soil) to perform their retaining function.

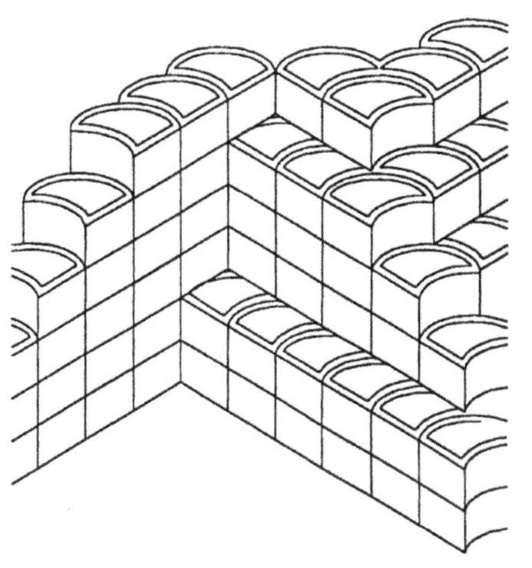

Fig 1.20 Precast concrete retainers

Activity 1.7

When driving around your area, make a point of stopping at a site where you think retaining walls are incorporated. Take photographs to assist in identifying these once you are back in class. Discuss with your lecturer or use the internet to establish the type and nature of these retaining structures. Why not build up a photo gallery of different retaining structures – some of these are really innovative and stunning. Some key words to use (if doing an internet search) are: retaining walls, reinforced concrete retaining walls, Loffelstein, CRB, etc. Also use the Concrete Manufacturers Association website at www.cma.org.za to assist in your search.

1.12 Construction plant

With the advances made in modern technology, the development of internal combustion engines has accelerated the mechanisation of construction activities. We rely on modern construction equipment (excavators, earthmovers and compaction plant) to perform tasks that normally would be undertaken by the available manpower. In some construction activities, it becomes more cost effective and less time consuming to use these machines.

Under this heading we will be examining some of the machines employed on site, their types, activities and their range of uses.

Construction Plant

Excavation plant	**Earth-moving plant**	**Compaction plant**
Skimmer	Bulldozer	Static weight roller
Face shovel	Angledozer	Vibratory roller
Backactor	Scrapers	Pnuematic roller
Dragline	Graders	Mini vibrating roller
Multi-purpose excavator	Tractor shovel	Vibrating plate
Trencher	Front-end loader	Impact plates

1.13 Excavation plant

The selection, management and maintenance of all construction plant is vitally important, especially in hauling and excavation activities.

Sometimes earth-moving plant can double-up as excavation plant and vice versa, so it is important to know:

- the site conditions;
- the activities that need to be carried out;
- the volume of work; and
- the economics of hiring equipment.

The final choice of plant to be used is usually left to the experience, familiarity with a particular manufacturer's products, availability and the personal preference of the site supervisor. Let's look at some examples of construction plant and their intended use.

Most excavating machines consist of a power unit (normally driven by a diesel engine) and an excavating attachment designed to perform a specific task or activity. Some of these machines are designed to carry out more than one activity simply by changing the excavating attachment.

1.13.1 Skimmer

A skimmer consists of a bucket sliding along a horizontal jib. The power unit is usually mounted on tracks. The bucket slides along the jib, digging away from the machine. Skimmers are used for very shallow excavations (usually up to a depth of 500 mm) and especially where level accuracy is required. To discharge the load, the boom or jib is raised and the power unit rotated until the raised bucket is over the haulage vehicle where it will dispose of its load.

1.13.2 Face shovel

This machine can be used as a loading shovel or for excavating into the face of an embankment or berm. It can be used in hard soil and in soft rock. The track-mounted power unit is available with a range of bucket sizes and capacities. The range for excavation is limited to between 300 mm below and 2.0 m above the machine level. The discharge operation is similar to that of the skimmer.

1.13.3 Backactor

This piece of plant is probably the most common form of excavating machinery. Contractors use it for excavating basements, trenches and pits. The power unit is normally mounted on tracks and the bucket size and capacity can vary. To discharge excavated material, the bucket is raised into a tucked position until the boom is positioned over the haulage vehicle and the material is emptied through the open front end. The haulage vehicle is usually parked alongside the excavation.

1.13.4 Dragline

This type of excavator is essentially a crane with a long jib or boom to which a drag bucket is attached. The drag bucket excavates loose and soft soil below the level of the machine. It is usually used in marshy areas where plant would become bogged down. These machines are also used for dredging rivers, canals and harbours. They are mounted on tracks and have a very long reach and a wide dumping range. Accurate use depends on the operator's skill. The bucket is cast out and drawn back by cables towards the power unit. Discharge of the collected spoil is similar to that of the backactor – through the open front end of the bucket. This machine can also be fitted with a grab bucket for excavating loose materials.

1.13.5 Multi-purpose excavator

These machines are based on a tractor power unit and are very popular among small- to medium-sized contractors because of their versatility.

They are popularly referred to as 'digger-loaders'. The tractor is usually a diesel-powered, wheeled vehicle, although tracks are available. It is fitted with a backacting bucket at the rear and a loading shovel at the front. During excavation, it is essential that the weight of the machine is removed from the axles. It achieves this through jacks at the rear of the power unit and inverting the bucket in the front.

1.13.6 Trencher

These machines are designed to excavate trenches of a constant width with considerable accuracy and speed. In most cases, they work in conjunction with laser technology that determines the required depth and level of the trench. Widths range from 250 to 450 mm and they can excavate depths of up to 4.0 m. Most trenchers work on a conveyor principle, with a series of small cutting buckets attached to two endless chains, supported by a boom that is lowered into the ground to the required depth. The spoil is normally deposited along the side of the trench, but some of the more sophisticated machines do the whole operation (trenching, pipe laying and backfilling) in one motion. These machines can excavate between 1.5 and 2.0 m per minute, depending on the ground conditions.

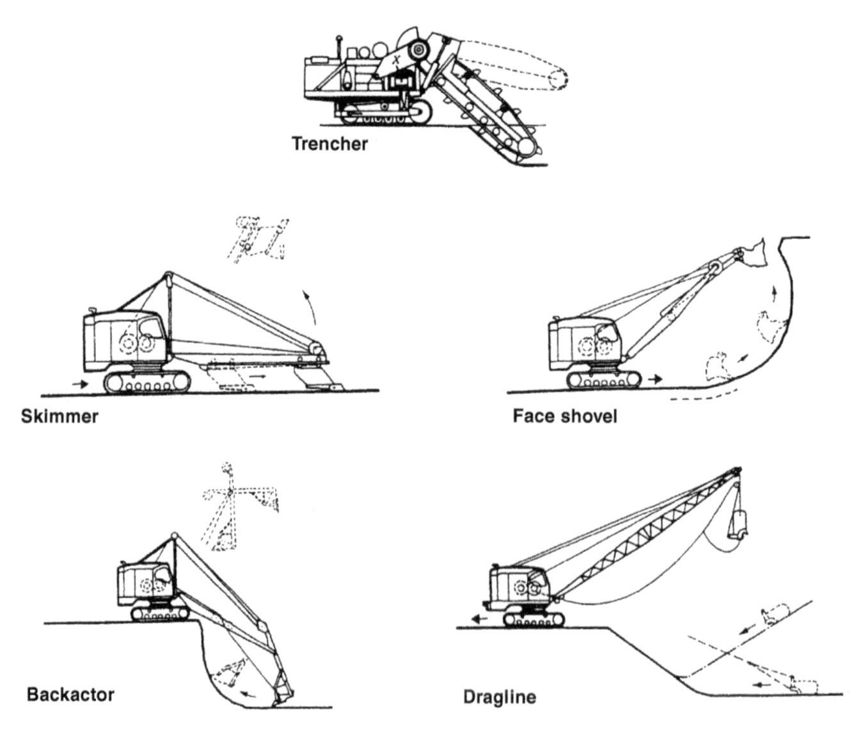

Fig 1.21 Excavation plant

1.14 Earth-moving plant

1.14.1 Bulldozers and angledozers

These machines are primarily high-powered tractors on tracks, fitted with a blade in the front for stripping and topsoil excavation up to a depth of approximately 400 mm (depending on the machine's specification). It does this by pushing the loosened material ahead of the machine. However, by changing the blade or mould board of the machine (adding a blade with teeth), it can be used in hard and rocky ground. This machine can be used for backfilling operations by simply setting the blade on the angledozer. A relatively level site can be obtained depending on the skill of the operator. These machines can be very large with blades of between 1.20 and 4.00 m in width and 600 mm to 1.20 m in height.

Bulldozers have other important uses, like ripping, grading and levelling, but sometimes these operations are done by plant specially equipped for this type of work.

Why are these machines mounted on crawler tracks instead of tyres?

Because of the nature of the work carried out, crawler tracks can't puncture, as is the case with normal tyre-operated plant. They also have the advantage of being less likely to become stuck (although this sometimes does happen) and tracks provide more traction when moving large amounts of soil. One of its disadvantages is that it is not as manoeuvrable as wheel-driven plant. It needs to be transported from site to site by a special transporting vehicle called a 'lowbed'.

Sometimes the tracks can break at one of the linkages, resulting in the machine being inoperable.

A **ripper** is a set of steel hooks (teeth) normally located at the rear of an earthmoving plant. It is used to rip up (plough) the ground.

1.14.2 Scrapers

This piece of machinery consists of a power unit and a scraper bowl. It is used to excavate and transport materials where surface stripping, site levelling and cut-and-fill activities are planned, particularly where large volumes of soil are involved. Scrapers are capable of producing a very smooth and accurately finished surface and come in three basic types:

- crawler-drawn;
- two axle; and
- three axle.

The design and basic operation of the scraper bowl is similar in all three types, consisting of a shaped bowl (capacities vary from 5 to 50 m^3) with a cutting edge that can be lowered to cut the top surface of the soil to a depth of 300 mm. As the bowl moves forward, the loosened earth is forced into the container. When full, the cutting edge is raised to seal the bowl. Sometimes it is necessary to use a bulldozer as a pusher to ensure a full load.

The bowl is emptied by raising the front apron and ejecting the collected spoil, or by raising the rear portion and spreading the collected spoil as the machine moves forward.

The crawler-drawn scraper is hooked behind a bulldozer and towed until the bowl is full. This is the slowest of the three, because it is governed by the speed of the bulldozer which is usually between 3 km/h when scraping and 8 km/h when towing. The three-axle scraper can reach speeds of up to 50 km/h. To achieve maximum output and efficiency, the following should be considered:

- When working in hard ground, the surface must be pre-broken by a ripper and cutting assisted by a pushing vehicle. Usually one bulldozer acting as a pusher can assist three scrapers.
- Where possible, the cutting operation should occur downhill, to take full advantage of the plant's weight.
- Haul roads should be kept smooth to enable the machine to reach maximum speed.
- Tyre pressures should be maintained to reduce forward rolling resistance.

1.14.3 Graders

These are popular machines used for road construction. They have adjustable blades that are fitted either at the front of the machine or slung under the centre of the machine's body. They are used to create the final finishing surfaces by 'cutting' or 'grading' the soil until the required levels are obtained. Great skill and experience is required of the operator. This is one of the more versatile machines available to the contractor as it can also grade inclined surfaces by moving the blade away from the body to the desired gradient.

1.14.4 Tractor shovel

Also called a loading shovel, this machine is basically a power unit on tracks with a hydraulically controlled bucket mounted at the front. Its primary function is to scoop up loose material or soil in the bucket, raise the loaded spoil and manoeuvre into a position to discharge the load into the back of a truck. Bucket sizes range from 0.5 to 4.0 m^3.

1.14.5 Front-end loaders

Front-end loaders, together with graders, are the most versatile machines available on a construction site. A front-end loader is a wheel-driven, diesel-powered unit that is fitted with a bucket at the front. It is more commonly used than the tractor shovel. Track-driven front-end loaders are also used, but are not as popular on site. Why do you think this is so?

Its operation is similar to that of the tractor shovel in that it uses its speed and momentum to force the bucket into the soil. Bucket capacities also vary depending on the machine's specification. They are often fitted with teeth instead of a straight cutting edge to enable excavating activities in harder soil.

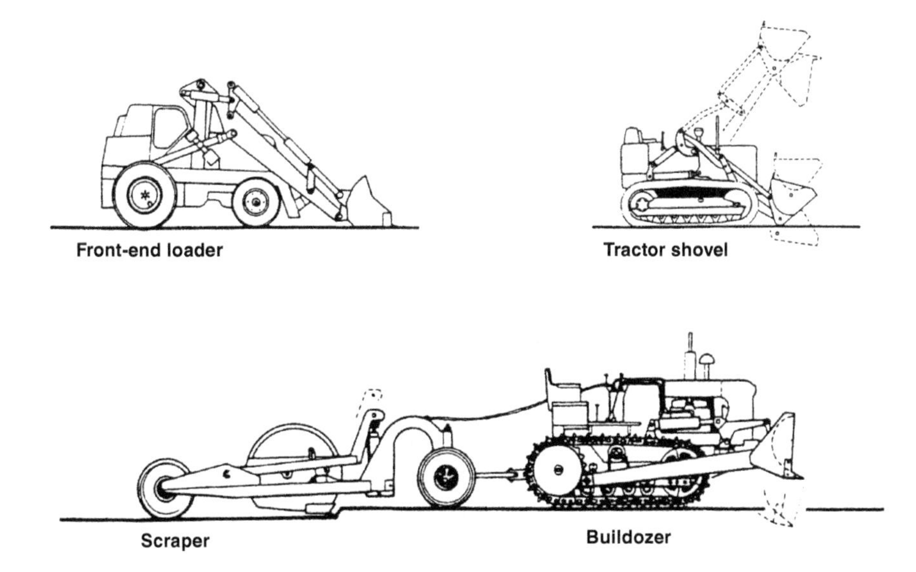

Fig 1.22 Earth-moving plant

1.15 Compaction plant

Compaction plant is designed to compact filling material and surface finishes, like premix for roads. Civil engineering contractors use large compaction equipment specifically for roadworks, whereas building contractors use smaller versions. The type of compaction obtained depends on the soil type. The actions can be described as vibratory, dead-weight and kneading (manipulation). Compaction plant is divided into static weight rollers, vibratory rollers and pneumatic rollers.

1.15.1 Static weight rollers

These rely upon the dead-weight of the machine to carry out the compaction. They are usually diesel powered and driven by a seated operator in a cab. They distribute dead-weight loads to the ground through two large-diameter steel wheels at the rear and one large steel drum in the front, which also acts as the steering wheel. Many of these rollers carry water tanks or have wheels that can be filled with water to add to the dead-weight. Some are also equipped with a steel tooth or scarifier at the rear of the vehicle for ripping up hard surfaces like roads.

1.15.2 Vibratory rollers

Vibratory rollers depend mainly on the vibrations produced by a petrol- or diesel-powered engine and can be self-propelled or towed behind another vehicle – for example, a bulldozer. The advantage of these machines over the static rollers is that they are lighter but produce the same or better compactive effort than their larger counterparts. They are also available with a combination of steel or rubber-tyred wheels. Vibration is imparted to the roll by means of a rotating off-centre weight situated inside the drum. Many models have the means to adjust both the frequency and the amplitude of the wheel vibration and it should be noted that high frequency with low amplitude is required when compacting a product like hot mix asphalt.

Vibratory rollers are particularly effective in granular soils.

Two types of vibratory rollers are:

- Smooth-wheeled roller (or static weight roller)
- Sheepsfoot roller (or kneading roller)

1.15.3 Pneumatic rollers

Pneumatic rollers knead or manipulate the soil structure. They can be distinguished from normal compaction plant by their smaller wheels that are made of rubber (usually seven wheels – three front and four rear) that are spaced in such a way that the front and rear treads interlock. These wheels can also move independently, thereby distributing the weight evenly on each tyre on uneven surfaces. They are self-propelled and using water for ballast can increase their weight.

Pneumatic-tyre rollers are used for rolling base courses and the final rolling of bituminous road surfaces.

The compactive effect of a roller is influenced by the following factors:

- static weight;
- the number of vibrating drums;
- frequency and amplitude of vibration;
- roller speed;
- the ratio between frame and drum weight;
- drum diameter;
- driven or non-driven drum; and
- the number of passes.

Refer to *Construction Materials*, where several compaction plant are discussed and illustrated.

Other forms of compaction equipment include:

- **Mini-vibrating rollers** (also referred to as bomags) that are hand operated.
- **Vibrating plates** consist of a mechanism supported on a steel plate that is allowed to vibrate, compacting the soil. This method is used on small patches and is more suitable for granular soils. Vibrating plates are manually operated by skilled labour. They are commonly known as plate compactors. You may have seen this type of compactor on a construction site or where road repairs are being undertaken.
- **Impact plates** for which kinetic energy is utilised by raising a heavy weight and allowing it to fall onto the surface.

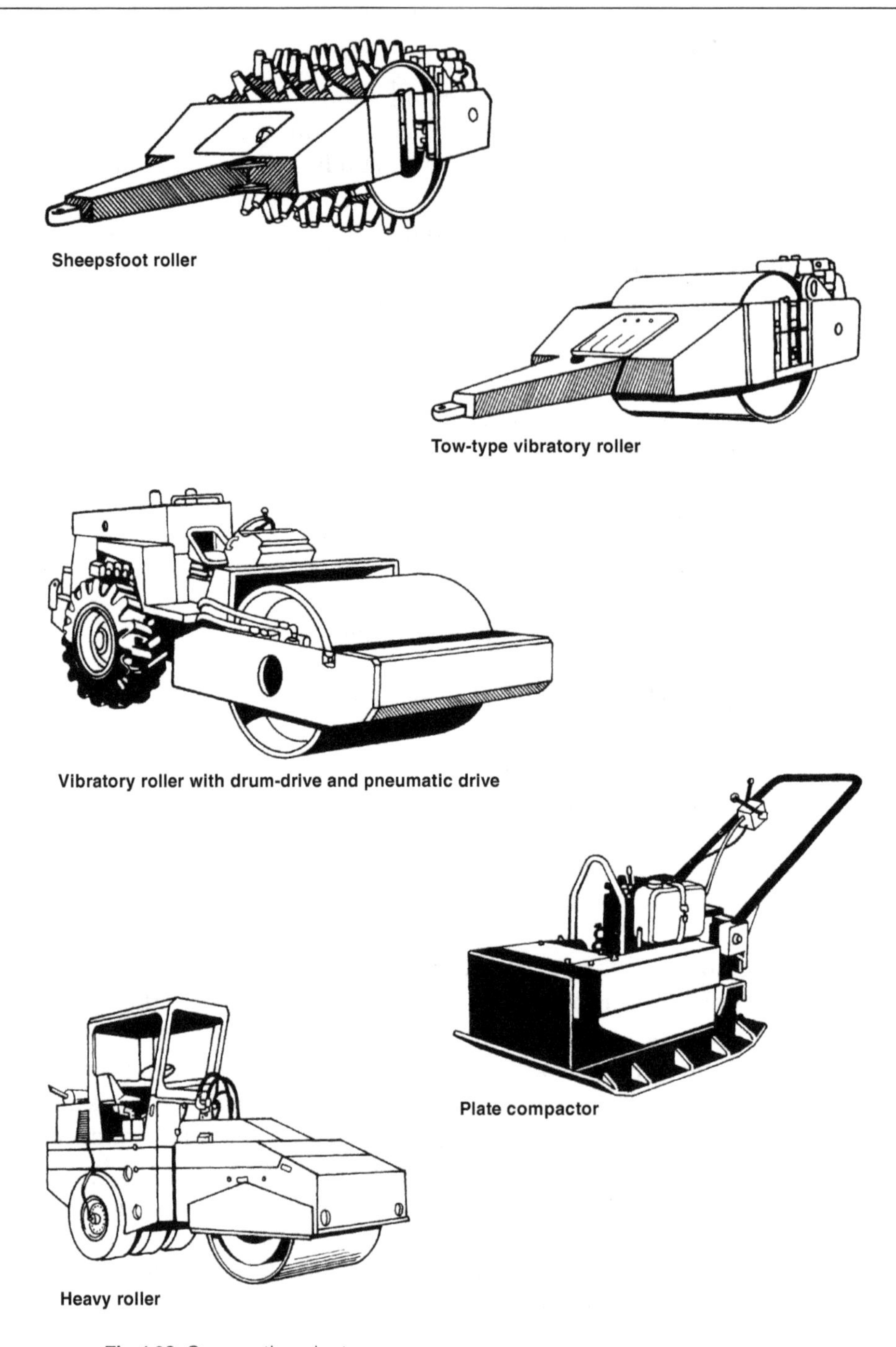

Fig 1.23 Compaction plant

1.15.4 Compactor zones of application

Activity 1.8

When next you pass a construction site, especially one where civil engineering works are being carried out, do the following:

1. Make a list of the construction plant you can identify at the site.
2. What function/s do they perform on site?
3. If possible, ask the site foreman about the operation of a particular type of plant.
4. Using the plant on the list you have compiled, obtain rates for each machine from hiring companies. Calculate how much it costs, on a daily basis, to operate each type of plant separately and then the total for all plant on site.

It is estimated that 35–45% of the cost of civil engineering projects is spent on materials with the rest being split up between labour and machinery.

South Africa is caught between the advancement of technology as well as the social obligation of providing work for the unemployed. It is therefore important to understand the processes involved with earthworks for excavations, utilising plant and safety (preventing sides of excavations caving in). There will always be a trade-off between man and machine but, for our purposes, it is important to understand the type and availability of construction plant, its work capacity and the range of functions it is able to perform. The decision regarding when to use man or machine is a debate left to higher powers.

With the potential increase in productivity using construction plant, it is also important to ensure that construction sites are kept safe. This is done by using accepted practices ensuring safety in the trenches. When carrying out excavation work on a congested site, the sides of excavations are frequently protected from caving in. When space is not limited, the sides are often kept open. It is important to know when to apply a particular method to ensure a safe construction site.

1.16 Summary

The purpose of this chapter was to:

1. Gain an understanding of what was meant by the term 'earthworks' and its relevance to civil engineering.
2. Explain some definitions used in earthwork applications.
3. Identify various foundation types.
4. Explain the meaning of timbering and its application in excavations.

5. Define retaining walls.
6. Identify and understand the uses of construction plant.

Self-evaluation 1

1. Complete the sentences:
   a. ____________ is the term used to describe all machinery used on a constructions site.
   b. Bearing capacity is measured in ______.
   c. ______ are box-like structures that can be sunk through ground or in water.
   d. ______ refers to the temporary supports for the sides of excavations.
   e. ______ refers to the soil below the topsoil.
   f. To provide a firm base, the base of the trench is covered with a layer of weak concrete, referred to as ______.
   g. Gravity retaining walls rely on their own ______ and the ______ on the underside of the wall to overcome the tendency to slide or overturn.
   h. A ______ consists of a bucket sliding along a horizontal jib.
   i. A ______ is a set of steel hooks normally located at the rear of an earthmoving plant.
   j. Graders are used to create the final finishing surfaces by ______ or ______ the soil until the required levels are obtained.
   k. The actions of compaction plant used on site can be described as ______, ______ and ______.
   l. ______ consists of a mechanism supported on a steel plate that is allowed to vibrate, compacting the soil.
2. State whether the following are true or false:
   a. Stockpiles refer to unusable material removed from site.
   b. Shallow foundations are deeper than 1.5 m below ground level.
   c. The density of concrete is 2 400 kg/m^3.
   d. Walings are vertical members running the length of a trench.
   e. Material too close to the edge of trenches may cause accidents.
   f. A diaphragm wall is a type of dividing or retaining wall.
   g. The basic function of a retaining wall is to retain soil at an angle greater than it would naturally assume.
   h. A face shovel can be used as a loading shovel or for excavating the face of an embankment.
   i. A scraper is a form of excavation plant.
   j. Static rollers depend on vibration for compaction of the soil.
   k. Low frequency and high amplitude is required when compacting a product like hot mix asphalt.
   l. A vibrating plate is also called a plate compactor. ➤

3. Answer the following short questions:
   a. What is settlement and how can it affect a structure?
   b. What is the function of foundations?
   c. How does one determine the nature and bearing capacity of a soil?
   d. When is it not necessary to provide timbering?
   e. Define a strut and the purpose it serves.
   f. Name the four basic methods for excavating basements.
   g. Why do we use construction plant today?
   h. What do you need to establish before hiring construction plant?
   i. Describe the factors that will influence the compactive effect of a roller.

Answers to self-evaluation 1

1. a. compaction plant
   b. kN/m^2
   c. caissons
   d. timbering
   e. subsoil
   f. blinder or blinding layer
   g. mass and friction
   h. skimmer
   i. ripper
   j. cutting, grading
   k. vibratory, deadweight and kneading
   l. vibrating plate
2. a. false
   b. false
   c. true
   d. false
   e. true
   f. true
   g. true
   h. true
   i. true
   j. false
   k. false
   l. true
3. a. see ref 1.2, page 4
   b. see ref 1.3, page 5
   c. see ref 1.5.1, page 10
   d. see ref 1.7, page 12
   e. see ref 1.8, page 13
   f. see ref 1.10, page 17

g. see ref 1.12, page 22
h. see ref 1.13, page 22
i. see ref 1.15.3, page 29

Chapter 2

Structures

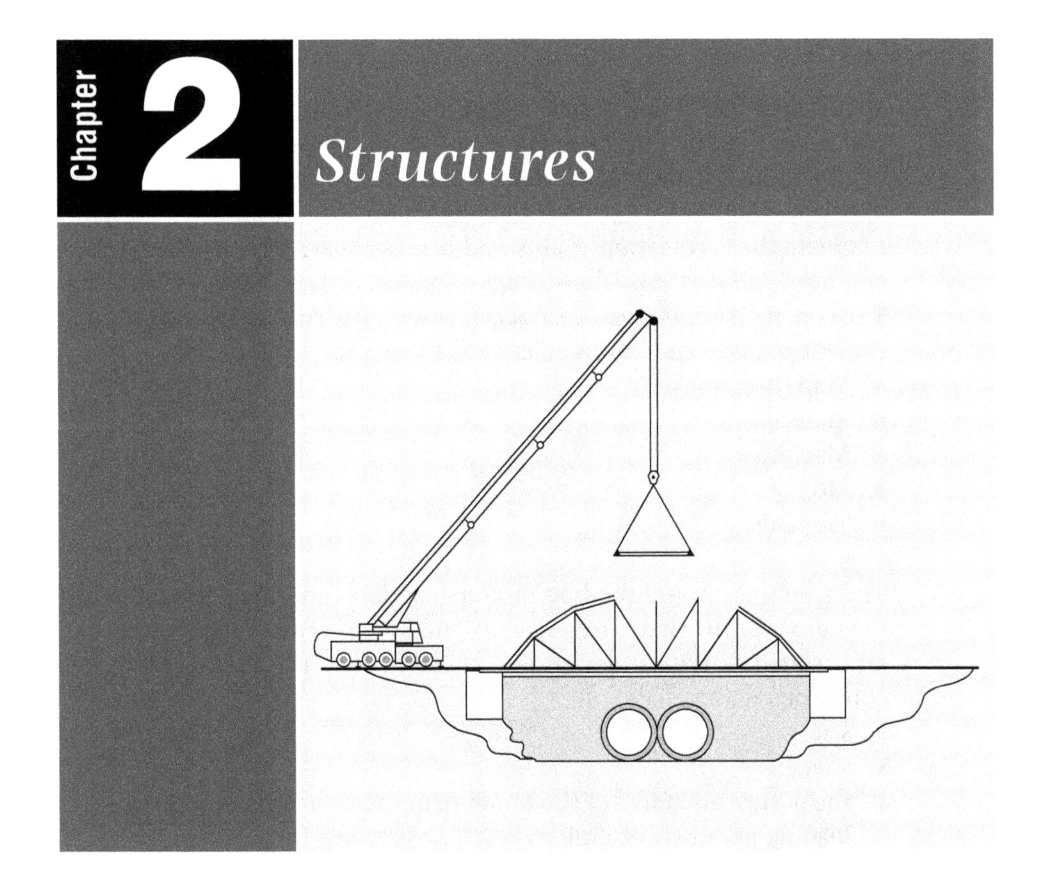

Outcomes

After studying this unit, you should be able to:

- Define structures as a general term
- Identify the functions of different professions
- Identify the different brick work bonding
- Know and understand the parts of a structure, i.e. substructure, superstructure and roof structure
- Discuss the application of scaffolding to structures
- Identify the various components of concrete and steel structures
- Discuss the use(s) of formwork as applied to concrete structures.

2.1 Introduction

'Structures', in broad terms, relate to buildings, bridges and dams, which serve many purposes. For example, the most important function of a building is to provide shelter. A shelter is basically a protection from the elements of nature and the function of a structure is to enclose space to create that protection. Dams and bridges also serve very important engineering functions: the storage of water or traversing an opening. The construction process is concerned with the rational and economic use of resources that many refer to as the **5 Ms**:

- Man (humans)
- Materials
- Machines
- Money
- Methods

The actual form and method of construction can differ depending on the nature of the structure. All construction processes follow two broad and related activities: **design** and **construction**. The design process is concerned with:

- size;
- shape;
- the nature and form of the building in relation to its function;
- loading parameters; and
- materials used.

The **construction process** is concerned with the nature and sequence of the operations – in other words, the method of construction. The design of the building, bridge or dam will, to an extent, determine the nature and sequence of building operations. One needs to consider the impact/s of the structure on the environment and vice versa. Environmental elements of weather, fire, temperature and sound must be considered when designing a structure. When it comes to the **design** of the building, two professional people are involved: an **architect** and a **structural engineer**.

The **architect** designs the structure to meet the basic needs of the client and performance standards, but also seeks architectural significance by making the structure aesthetically pleasing, environmentally sensitive and safe. This is done by using the materials available to define the building and give it its character.

Structural engineers, however, see to the structural requirements of the building, bridge or dam which involve the design of the foundations, beams, columns, slabs and roof structures within safety limits. The structural engineer's design must meet the architect's requirements.

A structure is divided into many parts:

- The **substructure** refers to the construction below ground level (some of this was covered in Chapter 1).
- Then there is the **superstructure** that relates to the actual building you see. This can consist of various forms – for example, brick, reinforced concrete, precast concrete, timber and structural steel frames.

There are some bridge arches cast in concrete that span wide gorges along the Garden Route. The most popular is probably the Bloukrans Bridge – it is 216 m above the Bloukrans River and the total span is 451 m across the gorge.

- In some instances, builders regard the roof as a structure in its own right (i.e. the roof structure).

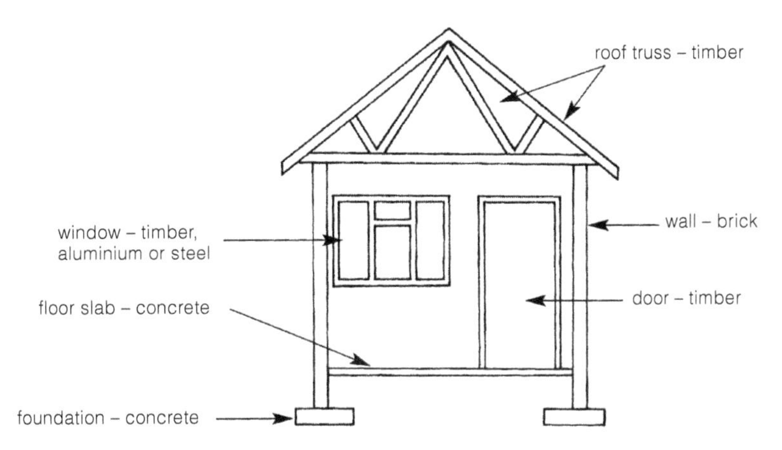

Fig 2.1 House construction

In this chapter, we will examine various forms of superstructure and their applications, along with the methods of construction.

2.2 Structural forms

There are various **structural forms** or **frames**, but the most common are:

- reinforced concrete buildings and bridges;
- structural steel buildings and bridges; and
- precast concrete buildings and bridges.

We will examine each of these in turn and discuss the materials and methods used in their construction.

2.2.1 Reinforced concrete frame

Refer to *Construction Materials* for more detailed notes about the properties of concrete.

Plain concrete is a mixture of cement, fine aggregate, coarse aggregate and water that undergoes a chemical reaction when mixed and sets as a rock-like mass. Concrete gradually increases in strength as curing takes place. A satisfactory **strength** is achieved after 28 days.

Does this mean that the concrete is not set before this time?

Concrete is usually set and able to withstand its load unsupported within 3 to 4 days after pouring. This is the reason why concrete formwork is removed during this time (and used elsewhere) as it does not serve much further purpose than to protect the concrete from drying out.

The concrete mass is strong in compression but not in tension. The strength of concrete depends on several factors:

- the type of cement used;
- the type and size of aggregates;
- clean mixing water;
- admixtures used; and
- the water/cement ratio.

Reinforced concrete has reinforcing steel in addition to the plain concrete mass. Steel or reinforcing steel is strong in tension. The combination of concrete and steel increases the ability of a member or members to carry additional loads. An advantage is that the size of the members can be reduced due to the extra strength provided by the steel. Reinforced concrete, however, is more expensive than ordinary mass concrete. Reinforcement is briefly covered in the *Construction Materials* handbook.

Reinforcement used in concrete must:

- have a high tensile strength;
- consist of a material that can be easily bent into any shape; and
- have a surface capable of developing an adequate bond between the concrete and the reinforcement, ensuring that the required tensile design strength is achieved.

Two types of steel are generally used as reinforcement in concrete – **mild steel** (R-bars) and **high yield steel** (Y-bars). The diameters can vary from 8 mm to 40 mm. The distinguishing feature between mild steel and high yield bars is that the latter (Y-bars) have longitudinal or transverse ribs for a better bond.

Reinforcement can be ordered directly from the supplier and either bent by the supplier into the desired shape or, alternatively, bending can be done on site using a bending machine. In Drawing II, the different bending shape codes and how to calculate cut-off lengths are investigated. Here is a sample to whet your appetite!

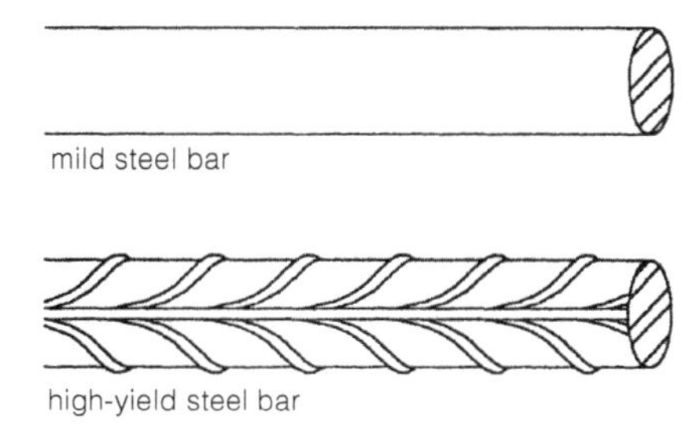

Fig 2.2 Steel reinforcement

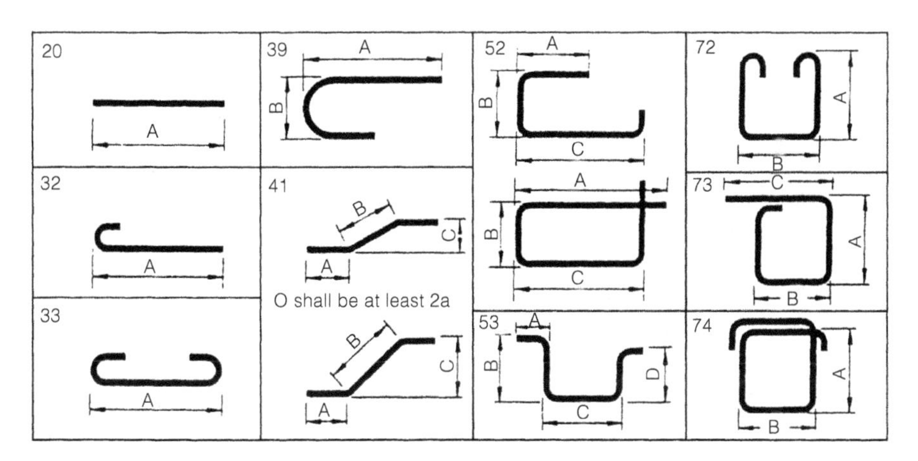

Fig 2.3 Shape codes

Diameter mm		6	4	10	12	16	20	25	32	40°	60°
Area mm²		28,3	50,3	78,5	113,1	201,1	314,2	490,9	804,2	1256,6	1963,5
Circum. mm		18,8	25,1	31,4	37,7	50,3	62,9	78,6	100,6	125,7	157,1
Mass kg/m		0,222	0,395	0,616	0,888	1,579	2,466	3,854	6,313	9,864	15,413
AREAS IN mm² GIVEN NUMBER OF BARS	1	28	50	79	113	201	314	491	804	1 267	1 984
	2	57	101	157	226	402	628	982	1 608	2 513	3 927
	3	85	151	235	339	503	943	1 473	2 413	3 770	5 891
	4	113	201	314	452	804	1 257	1 964	3 217	5 026	7 854
	5	141	251	392	565	1 005	1 571	2 454	4 021	6 283	9 817
	6	170	302	471	679	1 207	1 885	2 945	4 825	7 540	11 781
	7	198	352	549	792	1 408	2 199	3 436	5 629	8 798	13 744
	8	226	402	628	905	1 509	2 514	3 927	6 434	10 063	15 708
	9	255	453	706	1 018	1 810	2 828	4 418	7 236	11 309	17 671
	10	283	503	785	1 131	2 011	3 142	4 909	8 042	12 568	19 636
	11	311	553	863	1 244	2 212	3 456	5 400	8 546	13 823	21 598

Fig 2.4 Properties, areas and spacings of round bars (material reproduced from SANS 282: 2004)

Once the bars are cut and bent into the desired shape, the steel fixers can place and tie the bars in position, ready for the concrete pour.

Care must be taken to wire brush the bars to remove any loose rust that could result in poor bonding of the reinforcement and the concrete. Also check that there is no oil or grease on the bar.

The engineer or a representative – usually called a representative engineer or 'RE' – must physically check:

- that the correct bars (shapes) are used;
- the proper bar diameters are used;
- *the correct bar types are used;*
- that the bars are fixed as designed; and
- that the concrete cover to the reinforcement is as specified.

Bent reinforcement is fixed in the form of a cage for columns and beams and a mat for slabs and walls. Where the bars cross or intersect, they should be tied with soft wire or special wire clips, to prevent them from moving. **Spacer blocks** are added to maintain the cover (distance) between the reinforcement bars and the outside face of the structural member. This distance is necessary to prevent water and air seeping through the concrete onto the reinforcement and causing rust or corrosion. Spacer blocks are made from either concrete or plastic to the desired width needed for cover.

Design

A structural engineer or technician is usually asked to design reinforcing steel to carry the loads of a building or individual structural member. The designer will start by assessing the possible **self-weight and imposed loads** (dead and live loads) on a structural member and then calculate the reactions and effects these will have on the member. You will come across this in Theory of Structures at S2 level. Once the designer has established these effects, it is possible to determine where reinforcement is required and how much is needed. The calculations are based on the recommendations prescribed in the code of practice for the structural use of concrete – **SANS 10100-1**. This code identifies formulae that enable the designer to determine the area of steel required

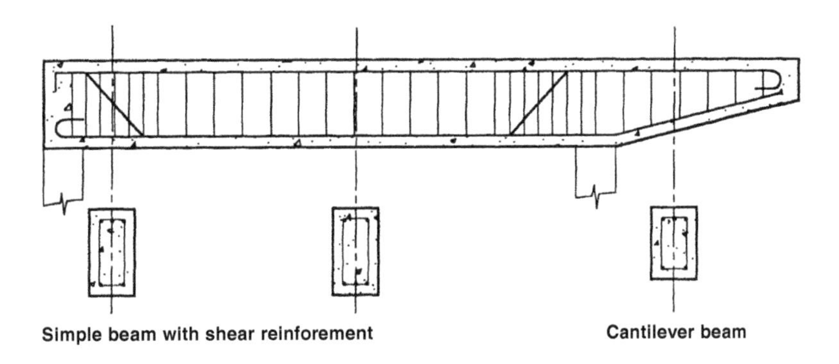

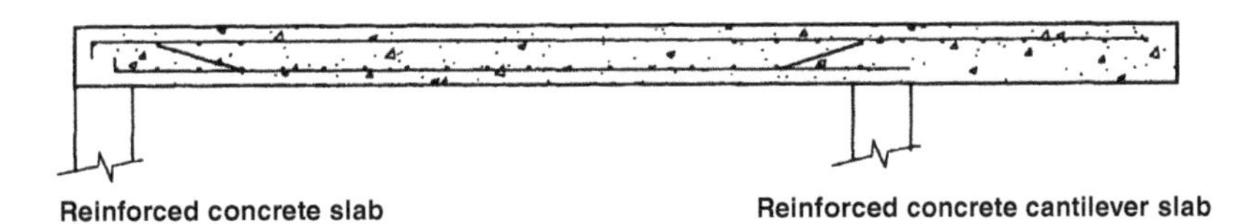

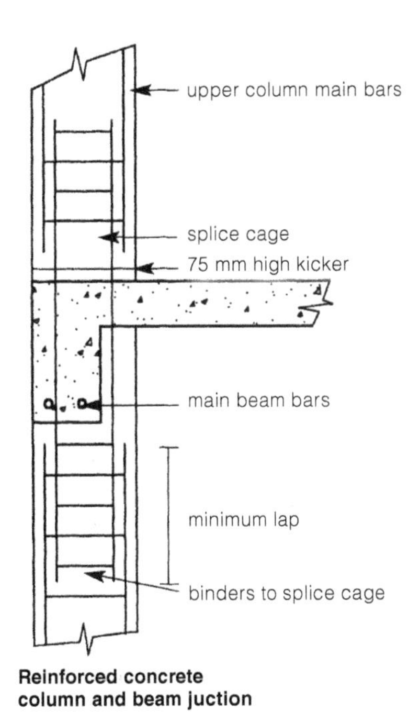

Fig 2.5 Typical examples of reinforced structural members

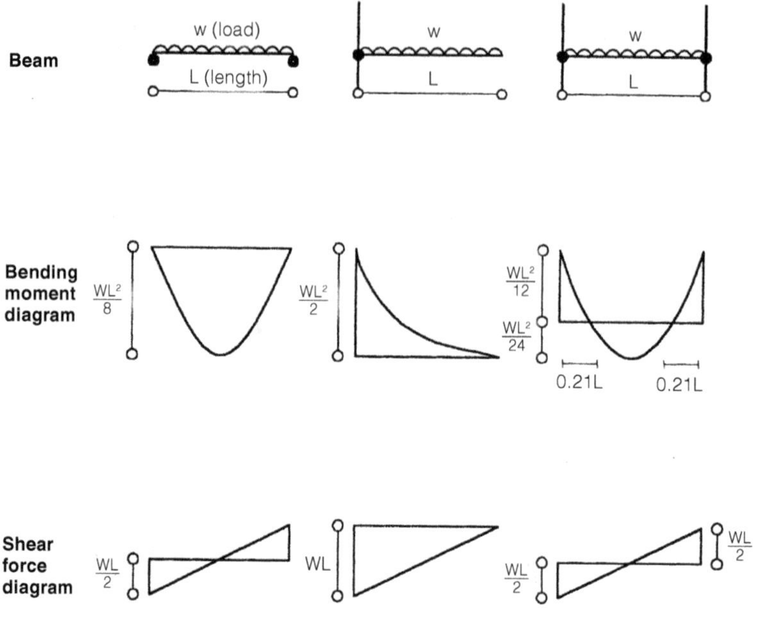

Fig 2.6 Calculation of structural loads

Once the reinforcement has been designed, detailed drawings are prepared and given to the contractor to construct the building. Copies are also sent to the suppliers for cutting and bending. Reinforcement drawings should specify the following table. For example, the specification:
8Y20-01-250B1

means
8 = total number of bars in the group
Y = high yield steel round bar
20 = diameter of bar in millimetres
01 = bar mark number
250 = spacing centre-to-centre in millimetres
B1 = location of reinforcing steel

When one prepares drawings showing reinforced concrete details, there should be a clear distinction between the lines drawn to represent the outline of the structure and those representing reinforcement. It is common practice to darken or use thicker lines to indicate the reinforcement.

Method of assembly before pouring

1. Any structural member, regardless of what it is (i.e. foundations, beams, columns or slabs), follows a similar approach.

2. Once the position of the structural member is finalised on site (usually by a surveyor), the formwork is erected around it. This must be securely fixed together as you do not want the formwork falling apart when the concrete is poured. The formwork and its fixings are usually done by an experienced carpenter or formwork operator.
3. After the 'box' around the structural member is secured, the reinforced steel is brought in and fixed in position. A steel fixer in conjunction with an engineer's representative ensure that the reinforcement is properly fixed and in its correct position. Drawings are used and interpreted in order to position the reinforcement together with a measuring tape and tie-wires to ensure that the reinforcement does not move when the concrete is poured.
4. An important part before fixing the reinforcement is to ensure that the correct cover is observed between the reinforcement and the outer surface of the concrete. Usually 'spacer' or 'spacing' blocks are fixed to the outer layer of reinforcement and the formwork to ensure that the correct 'cover' is achieved.

We need to have cover between the surface of concrete and the reinforcement because steel is subject to corrosion and rust and when this happens, it compromises the strength of the reinforcement, reducing its strength.

5. During the time that the reinforcement and formwork is being checked for correctness and accuracy, the concrete is ordered. Usually the concrete arrives in 6 m^3 ready-mixed concrete trucks but, on large projects, a batch plant may be erected to supply the concrete. Before the concrete pour starts, the area inside the formwork is cleaned and all debris is removed so that it does not contaminate the concrete.
6. Depending on where the structure is, the concrete can be moved via wheelbarrows, directly from the truck (or extended chute), or pumped using a concrete pump or a bucket attached to a tower crane.
7. It is important to ensure that the concrete is well compacted and the voids are reduced – this is achieved by using a vibratory poker to expel the air and 'compact' the concrete mix.
8. Once the concrete pour is complete, the concrete is smoothed and protected against excessive moisture loss.
9. Take care to take samples and undertake quality testing to ensure compliance.

2.2.2 Reinforced concrete beams

The most common forms of beam are the simply supported, continuously supported, and cantilever beams. The complexity of the design and the loading will determine the reinforcement necessary.

Generally, when a simply supported beam is loaded, the fibres in the beam will stretch or compress until the ultimate strength is reached, then cracking and subsequent failure occurs. The fibres at the bottom of the beam will experience the stretch (called **tensile stress**). We commonly refer to the bottom of the beam as being in tension (where tensile stresses are active). The top fibres will be compressed or in compression (where **compressive stresses** are active). The correct design of the reinforced concrete beam will ensure that there is sufficient strength to resist both compressive and tensile stresses.

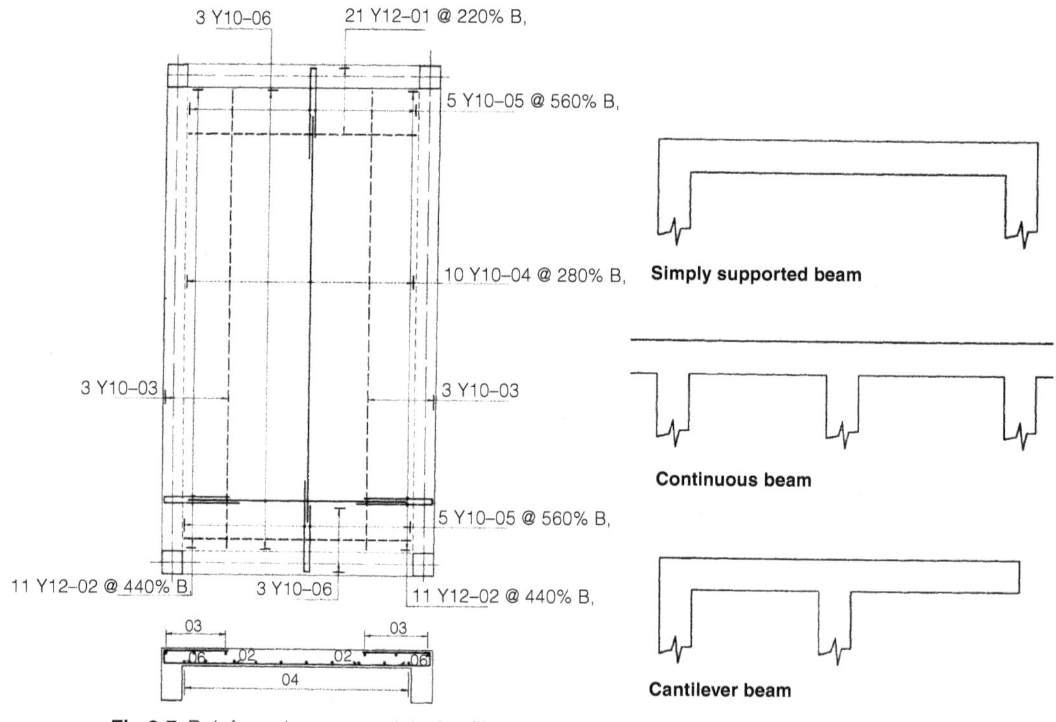

Fig 2.7 Reinforced concrete slab detailing

Fig 2.8 Typical beam sections

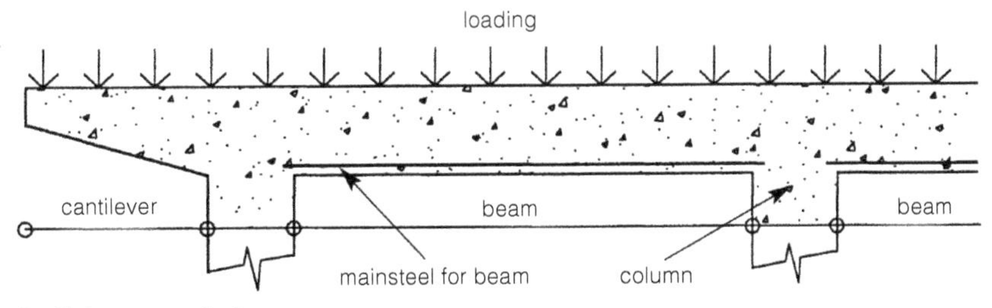

Continious span design

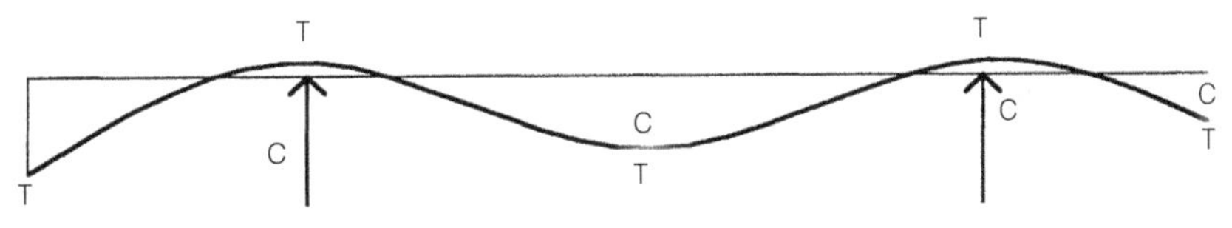

Tension and compression zones

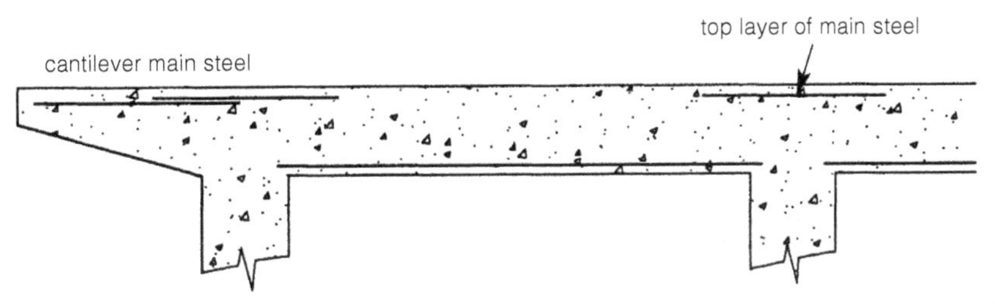

Reinforcement in tensile zones

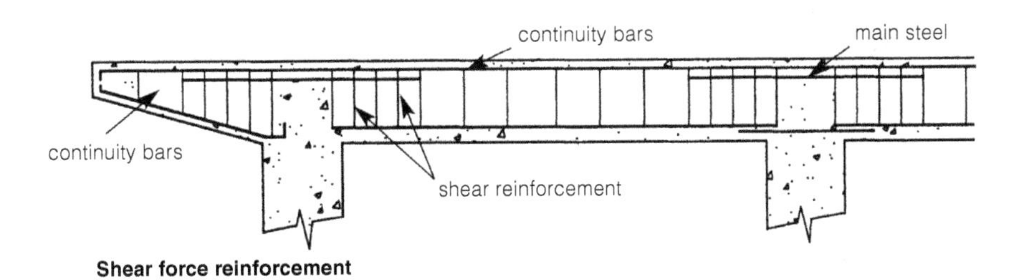

Shear force reinforcement

Fig 2.9 Beam reinforcement

Activity 2.1

Take a normal eraser and, keeping it in both hands, bend it downwards. Notice what happens to the fibres at the bottom of the eraser (they are stretching). This resembles the tensile stresses of a beam. Now look at the fibres at the top of the eraser that are being forced together. This resembles the compression effect in the beam.

Now take two fairly thick sewing needles and slowly push each needle into the lower half of the eraser along its length. Repeat the experiment. See if more effort is now needed to bend the eraser. Explain your findings.

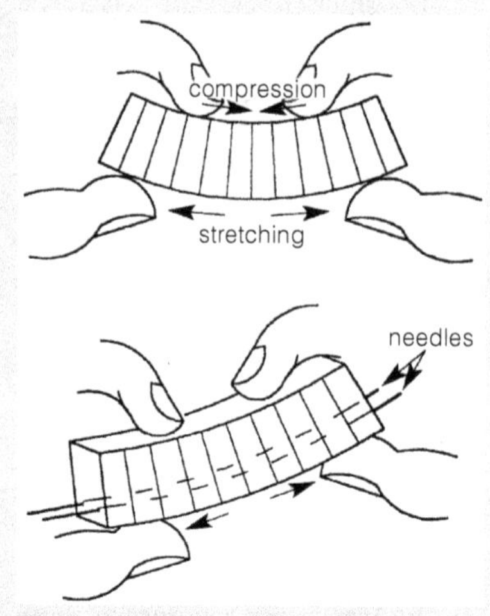

Fig 2.10 Stresses of a beam

Activity 2.2

Get into groups of four and try the following experiment. Each group is given:

- a knife;
- a length of 300 mm × 70 mm × 3 mm balsa wood; and
- 150 mm masking tape.

Your group must design a beam of any shape using only the materials supplied. The beam will be tested by applying loads at the centre of its span until ultimate strength is reached or failure occurs. The beam must span a length of at least 250 mm. No laminated beams are acceptable.

Take about 30 minutes to complete this activity and then start testing. See which group has the best or strongest design.

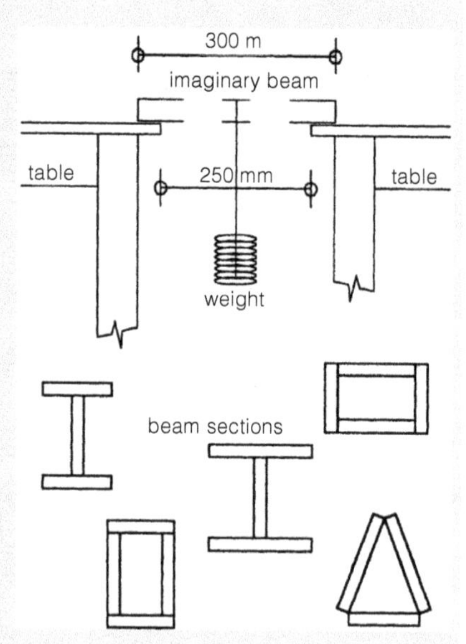

Fig 2.11 Typical beam shapes

2.2.3 Reinforced concrete columns

A **column** is a vertical member that carries the beam and floor loads to the foundation.

A column is a compression member. In other words, the fibres in the columns are being forced together. It is like taking a cigarette and placing it upright on a table. Place your finger on the end of the cigarette and push it down towards the table. Ultimately, the cigarette will become smaller until it eventually collapses or breaks. A column may go through a similar experience and buckle, collapse or deflect (bend).

Because concrete is strong in compression, it may be concluded that, provided the compression strength of the concrete is not exceeded, no reinforcement is needed. For this to hold true:

- the loading must be axial;
- the column must be defined as short; and
- the cross section of the column must be large (wide) relative to the length.

It is very rarely that these conditions occur in large buildings and bridges. Consequently, bending is induced and there is a need for reinforcement to provide tensile strength. **Bending** may be induced by:

- reaction of the beams upon the columns;
- wind loading; or
- earthquake loading.

When bending becomes excessive, concrete and steel columns will buckle. Slender columns tend to fail due to buckling. It is therefore essential that, when doing structural designs, you consider the buckling effect of columns.

Look at figure 3.25 in the *Drawing* handbook to see how concrete columns are detailed.

Activity 2.3

To demonstrate the buckling effect, do the following simple exercise. Take a 300 mm plastic ruler and place it on its shortest end on a hard surface (e.g. a table). The ruler is now perpendicular to the surface of the table. Apply an axial force to the other end of the ruler using your finger to press down towards the table. If you press hard enough, the ruler will start bending near the middle. This bending is referred to as buckling. Once buckling occurs, the member will go into failure and collapse, as it can no longer withstand the axial force.

In most cases (particularly with square or rectangular columns) the minimum number of main bars is four (and six for circular columns). To prevent these long bars from buckling, links or stirrups are used as restraints. **Stirrups** are also used to prevent cracking under high loads. Where the junction between column and beam bars occurs, there could be a clash of steel since bars from the beam could be in the same plane as bars in the column. To avoid this situation, one group of bars must be bent into another plane (usually the column bars).

2.2.4 Reinforced concrete slabs

A reinforced concrete slab will act in exactly the same manner as a reinforced concrete beam and is therefore designed in the same manner. The same parameters as for other structural elements will be investigated:

- loading;
- bending moments;
- shear forces; and
- reinforcement requirements.

In most reinforced slabs, the reinforcement is assembled to form a continuous mat.

Shear stresses is another important consideration when designing structural members. They tend to cause sliding between adjacent sections.

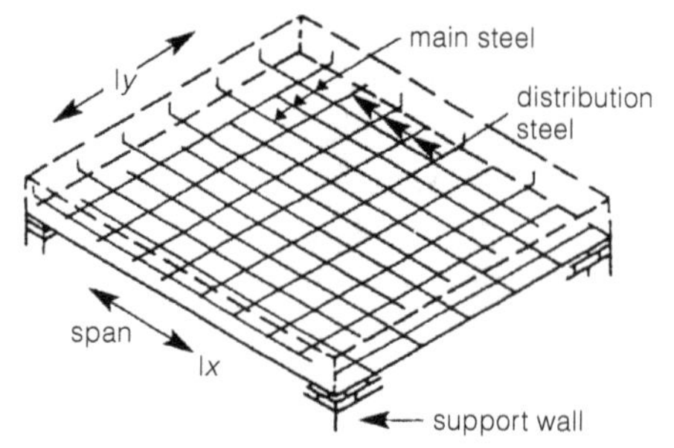

Fig 2.12 Slab reinforcement

Generally there are two types of stresses:

- **Vertical shear stresses**, which occur near supports as a result of heavy loads tending to cause the central section of a member to slide vertically downwards.

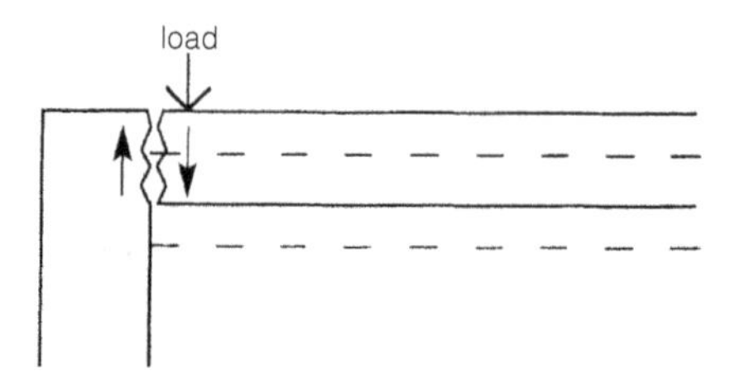

Fig 2.13 Vertical shear force

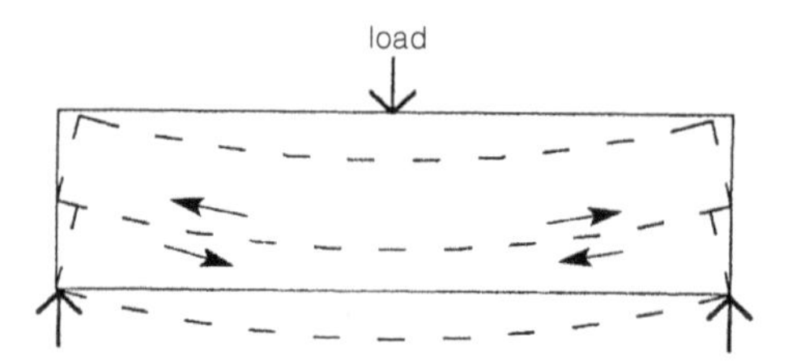

Fig 2.14 Horizontal shear force

- **Horizontal shear stresses**, that result from the tendency of a beam to bend under load and split into horizontal laminations.

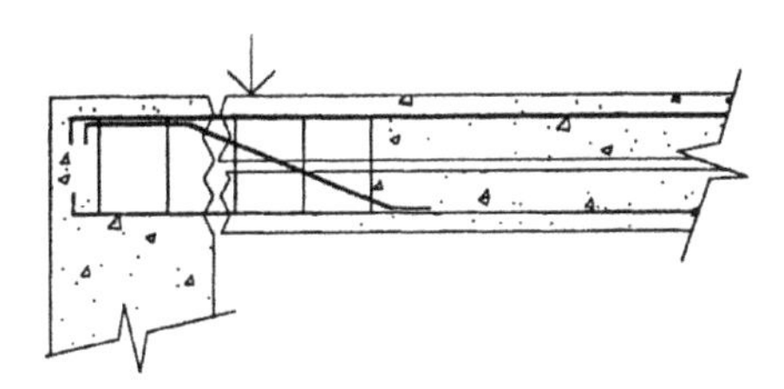

Fig 2.15 Steel reinforcement to counteract shear forces

At the ends of heavily loaded members, such as beams, a combination of vertical and horizontal shears results in diagonal stresses. To prevent diagonal cracking at the ends of beams or adjacent to supports, it is often necessary to either bend up some of the tensile reinforcement or to use stirrups.

2.3 Structural steel forms

Using structural steel in building construction began at the beginning of the 20th century and gained popularity in the construction of small- to medium-span industrial buildings. As with reinforced concrete, the elements of the building (the columns and beams) remain the same, but are replaced by structural steel members. Structural steel has been used and is being used in bridge construction (see Chapter 5), portal frames, trusses, girders, etc.

A **portal frame structure** consists of a series of steel beams and columns that are connected in such a way that they form an arch. This creates large, clear working areas inside a building.

Structural steel frames are easily assembled and erected. They are relatively expensive compared to reinforced concrete and are not very labour intensive. The columns and beams are welded or bolted together.

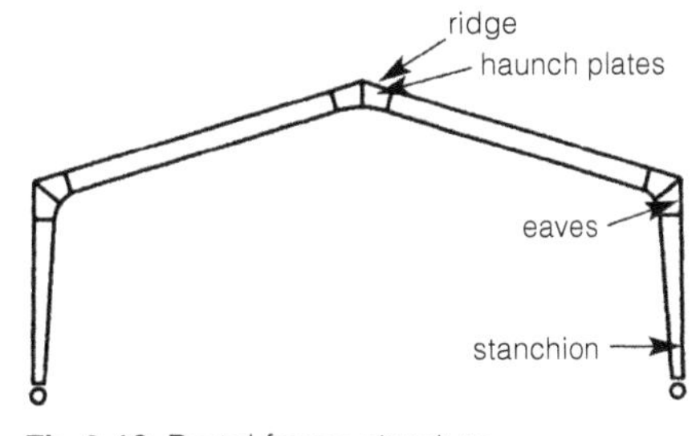

Fig 2.16 Portal frame structure

The design of a structural steel building is normally placed in the hands of an experienced structural engineer or technician who calculates the loadings, stresses and reactions as for reinforced concrete.

Once this is done, the specialist will select a standard steel member with section properties that meet design requirements.

The corner connections are stiffened with haunch plates. To make transportation easier, the members are fabricated in manageable sections. On site, they are assembled through strategic connection points to facilitate erection of the frame.

Several frames are usually connected to form a building, using beams and cross bracing to provide lateral stability.

Portal frames can also be made using precast concrete elements.

It is the building contractor's function to ensure that the base foundations are in the correct position and at the correct levels, with the necessary holding-down fixing bolts. In Drawing II you will be expected to complete a structural steel drawing showing all details and sections.

As for a reinforced concrete design, structural steel designs must be done according to a code of practice – **SANS 10162**. In conjunction with this code, a *Handbook on Structural Steelwork* is used to determine the section types. There are various section types:

- universal beams (or I-sections – parallel flange);
- universal columns (or H-sections – parallel flange);
- rolled steel joists (taper flange);
- angles (equal leg and unequal leg);
- channels (taper flange and parallel flange); and
- T-sections.

2.3.1 Universal beams (UB)

A range of universal beam sections is available and described by their size (in millimetres) and mass (in kilograms) per metre run. A beam or rafter is also specified in the following manner. For example:
356 × 171 × 45 I

means
356 = depth of beam/rafter
171 = flange width
45 = mass of the section in kg/m
I = I-section
(All dimensions are in millimetres.)

The value 356 × 171 × 45 I is the designation only. The actual dimensions must be found in the structural steel tables. To see the actual dimensions of this section, look at the extract from the steel tables shown below.

I-sections (parallel flange)

Dimensions and properties

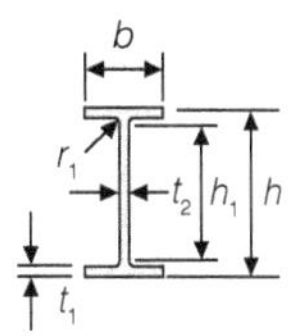

Designation	m	h	b	tw	tf	rl	hw	A
mm × mm × kg/m	kg/m	mm	mm	mm	mm	mm	mm	10^3 mm^2
356 × 171 × 45	44.8	352.0	171.0	6.9	9.7	10.2	312	5.70
51	50.7	355.6	171.5	7.3	11.5	10.2	312	6.46
57	56.7	358.6	172.1	8.0	13.0	10.2	312	7.22
67	67.2	364.0	173.2	9.1	15.7	10.2	312	8.55
406 × 140 × 39	38.6	397.3	141.8	6.3	8.6	10.2	360	4.92

Fig 2.17 Universal beam sections

2.3.2 Universal columns (UC)

Universal columns are similar to beams and are described in the same manner. It is possible to design a column section to act as a beam and vice versa. The specification is similar to that for a universal beam.

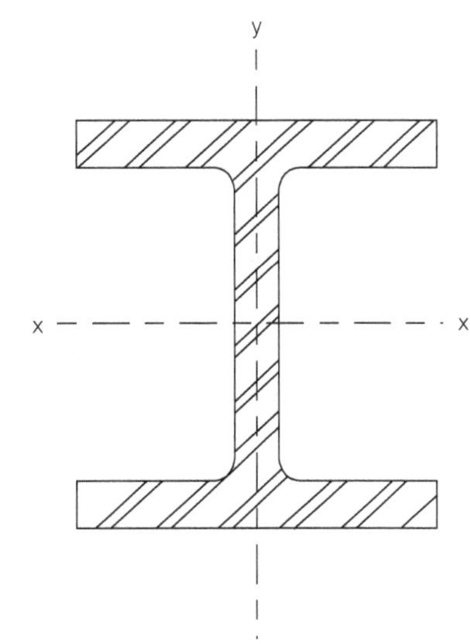

Fig 2.18 Universal column

2.3.3 Rolled steel joists

These are a range of small-size beams that have tapered flanges and are used for lintels and small frames around openings.

2.3.4 Angles

Angles are light framing and bracing sections with perpendicular legs. The legs can be equal or unequal and the sections are described by their leg length (in millimetres) and the nominal thickness of the flange. For example, 60 × 60 × 5 means:

60 × 60 = equal leg length of 60 mm

5 = leg thickness of 5 mm

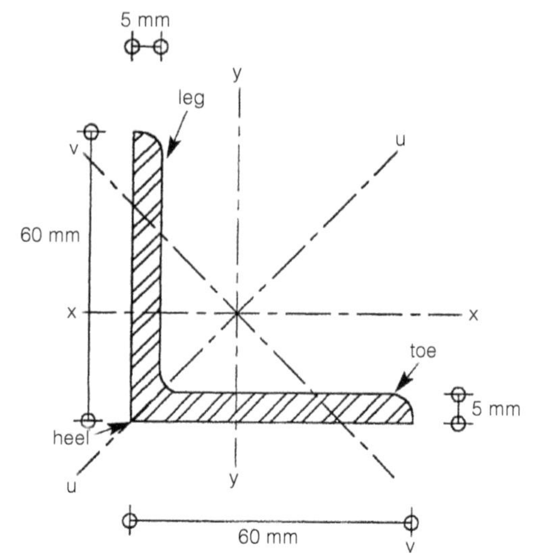

Fig 2.19 Angle section

2.3.5 Channels

Channels can be used for trimming and bracing members or as substitute joists. They are specified according to their nominal overall dimensions and their mass per metre run. To get an idea of the sizes used for a typical channel, look at the extract from the steel tables below.

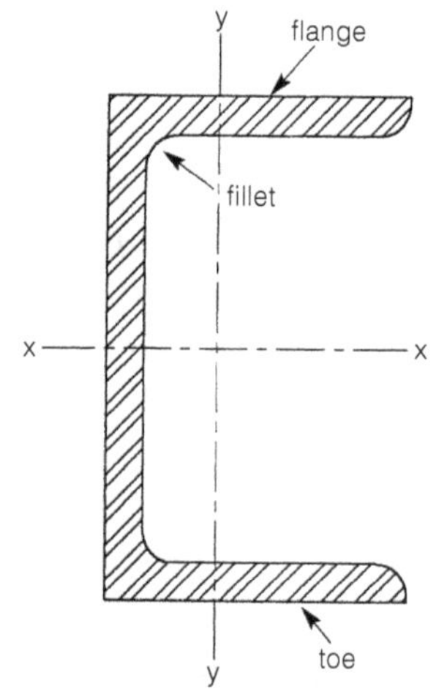

Fig 2.20 Channel section

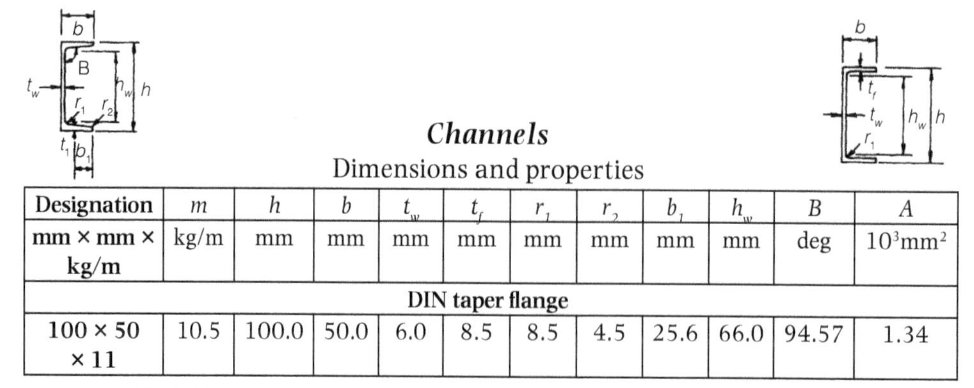

Channels

Dimensions and properties

Designation	m	h	b	t_w	t_f	r_1	r_2	b_1	h_w	B	A
mm × mm × kg/m	kg/m	mm	mm	mm	mm	mm	mm	mm	mm	deg	$10^3 mm^2$
DIN taper flange											
100 × 50 × 11	10.5	100.0	50.0	6.0	8.5	8.5	4.5	25.6	66.0	94.57	1.34

Fig 2.21 Channel section properties

2.3.6 T-sections

T-sections are useful for the same purpose as angles. They can be cut from a standard universal beam or column. They are specified according to their nominal overall width and depth, and their mass per metre run.

Look at the extract from the structural steel tables below so that you can see how the typical dimensions are listed. Note that this is only one of the many T-sections available and should therefore not be viewed as standard.

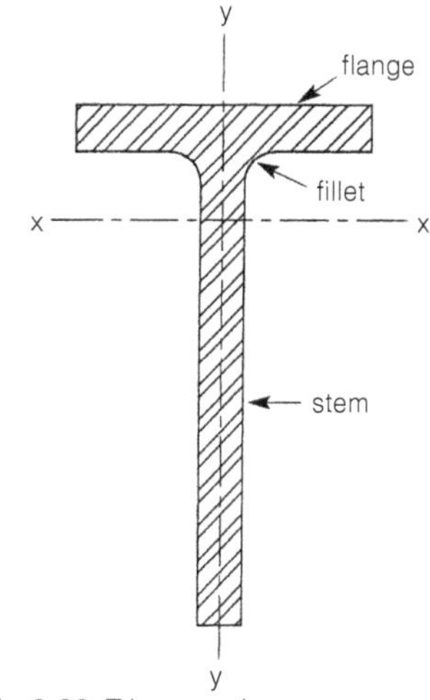

Fig 2.22 T-bar section

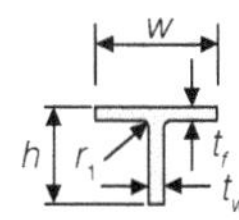

Structural tees cut from H-sections
Dimensions and properties

Designation	m	Cut from I-section	*h*	*b*	t_w	t_f	r_1
h × *b*		Designation					
mm	kg/m	mm × mm × kg/m	mm	mm	mm	mm	mm
76 × 153	11.7	152 × 152 × 23	76.2	152.4	6.1	6.8	7.6
	15.1	30	78.7	152.9	6.6	9.4	7.6
	18.5	37	80.9	154.4	8.1	11.5	7.6
	23.1	203 × 203 × 46	101.6	203.2	7.3	11	10.2
	26	52	103.1	203.9	8	12.5	10.2
	29.8	60	104.8	205.2	9.3	14.2	10.2
	35.7	71	108	206.2	10.3	17.3	10.2
	43.2	86	111.1	208.8	13	20.5	10.2
	36.5	254 × 254 × 73	127	254	8.6	14.2	12.7
	44.6	89	130.2	255.9	10.5	17.3	12.7
	53.6	107	133.4	258.3	13	20.5	12.7
	66.2	132	138.2	261	15.6	25.3	12.7
	48.4	305 × 305 × 97	153.9	304.8	9.9	15.4	15.2
	58.8	118	157.2	306.8	11.9	18.7	15.2
	68.4	137	160.3	308.7	13.8	21.7	15.2
	78.8	158	163.6	310.6	15.7	25	15.2

Fig 2.23 Structural tees cut from H-sections

2.3.7 Connections

When using structural steel in construction, the foundation is usually a reinforced concrete base or pad footing, designed to carry the loads transferred.

Three types of connection can be used to connect the steel frame to the base:

1. **Pin or hinge connection:** Special bearing plates are held down with bolts driven into the concrete foundation.
2. **Base-plate connections:** There are two forms of base-plate connection: slab and gusset.
3. **Pocket connection:** The foot of the supporting member is inserted and grouted into a pocket formed in the concrete foundation.

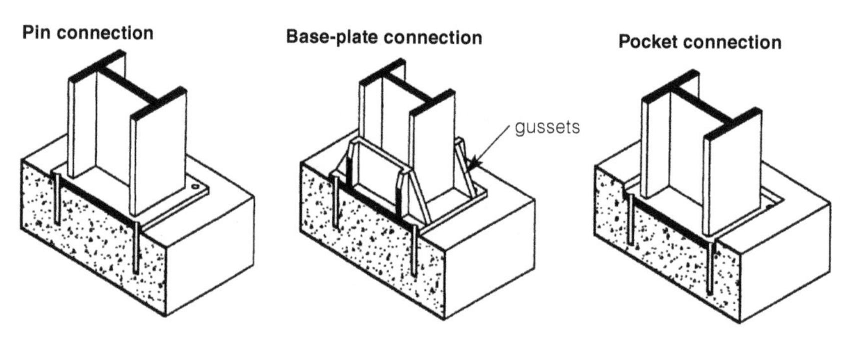

Fig 2.24 Steel connections

Connections in structural steelwork are classified as either shop connections or site connections and can be made using bolts or rivets or by welding. Shop connections are done in a workshop before delivery to site. Site connections are done as the structure is being erected.

Bolts

Black bolts are the cheapest bolts available, and are either hot or cold forged. Problems in using these bolts relate to shearing as the allowable shear stresses of this bolt are low. These bolts are mainly used for end connections or in areas where seating brackets are used to resist shear forces.

Rivets are made of mild steel but are being superseded by either bolted or welded connections. They are available in a variety of head shapes from semi-circular to countersunk.

High strength friction bolts are manufactured from high tensile steel and are used to replace black bolts and rivets. Typical bolt sizes range from 12 to 36 mm diameter, with lengths of between 40 and 500 mm.

Welding

Primarily considered a shop connection since the cost and the need for inspection generally makes this method uneconomical for site connections.

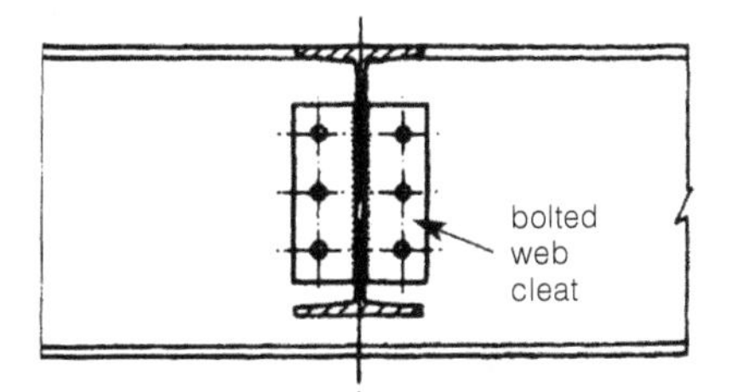

Electric arc and oxy-acetylene welding are generally used. With the latter, a metal filler rod is held in the flame and the molten metal from the filler rod fuses the surfaces together.

Welds are classified as either fillet or butt welds. **Fillet welds** are used at the edges and ends of members and form a triangular fillet of welding material. **Butt welds** are used on chamfered end-to-end connections. Refer to the *Drawing* handbook for more information.

Beam to column connections can be designed as simple connections, where the whole of the load is transmitted to the column through a seating cleat. A rigid connection detail is made by welding the beam directly to the column, making it more economical.

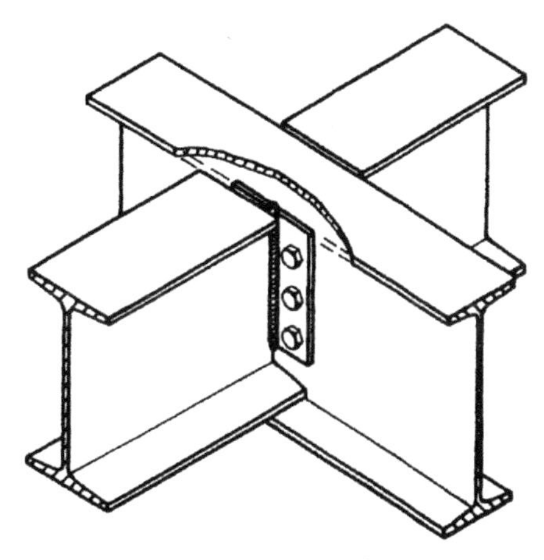

Fig 2.25 Steel beam connection

Beam to beam connections depend on the relative depths of the beam concerned. Deep beams receiving small secondary beams can have an angle connection, whereas others can be connected using web cleats.

When you reach S3 and S4, you will study the elements of structured steel, from design through to materials.

Activity 2.4

When next you go on a site, visit to where they are erecting a structural steel frame, make a note of the types of sections used and the way in which they are fixed or secured. Try to take some photographs and, in groups, create a large poster identifying these sections or procedures.

Ask the site foreman if there are some off-cuts of the various sections that you can bring to the class for reference.

Remember to ask first before you remove anything from a site.

2.4 Precast concrete forms

The overall concept of precast concrete is the same as those of reinforced concrete and structural steel. Most precast concrete frames are produced as part of a 'system' building, where the various parts or elements are slotted together to complete the product. The most common product is a precast concrete garage that is designed, fabricated and delivered to site in one complete unit. Modern school buildings are erected using precast concrete sections that are assembled and secured on site, to form a complete building in a fairly short time. Precast concrete elements are also used in some bridges.

Precast concrete, as with other forms of concrete, is comprised the basic ingredients; concrete, aggregate and water. However, the unit manufacturing process is different. The mixture is made, placed into moulds, compacted and cured. It is important that the moulds are not removed until the concrete has gained sufficient compressive strength. The curing standards that apply to reinforced concrete are also applied to precast concrete.

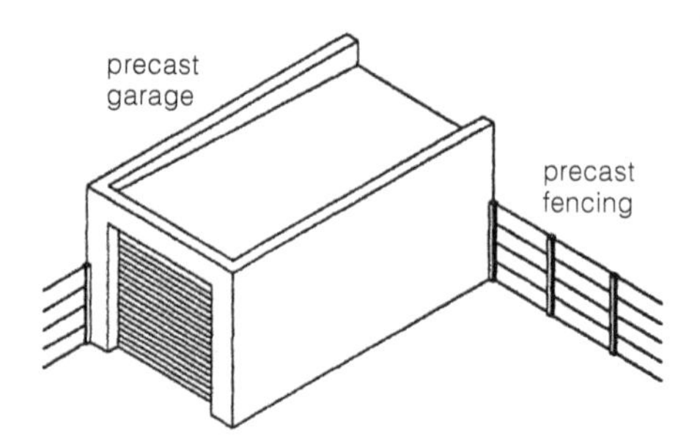

Fig 2.26 Precast concrete forms

Precast concrete products include:

- Septic tanks
- Manholes
- Box culverts
- Channels
- Prestressed bridge beams

- Bridges
- Concrete road barriers, etc.

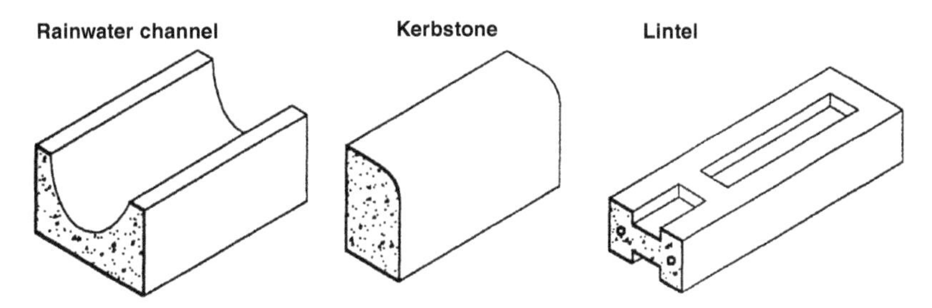

Fig 2.27 Precast concrete products

Precast concrete has both advantages and disadvantages over the more conventional construction methods.

Advantages

- The mixing, placing and curing of the concrete is carried out at the factory (under controlled conditions) resulting in good quality products.
- This off-site activity makes more space available on the site for storing other materials like cement, sand and aggregates.
- Mass production reduces cost.
- The complete sections can be assembled on site in almost any weather, minimising delays.
- The assembling is done by semi-skilled workers, which releases skilled workers for other activities and allows for quick and easy training of unskilled labour.

Disadvantages

- Because of the nature of its components, the designs of precast buildings are relatively inflexible compared to, for example, reinforced concrete designs. Precast concrete forms are made to precise measurements and, should there be a discrepancy in the setting out of the construction works, major problems could result on site. Reinforced concrete is more adaptable to these inaccuracies.
- Large cranes are needed to lift sections into place and, if the site is very limited in space, manoeuvrability will be problematic.
- Structural connections between the precast concrete units can present both design and contractual problems.
- Job planning and programming may be restricted due to the specialised transportation and delivery requirements of precast concrete units.

Activity 2.5

Look in books, magazines and newspapers, and see if you can come up with more examples of precast concrete units.

2.5 Scaffolding

Scaffolding is a temporary structure from which people can gain access to high-level working areas to carry out construction operations. For example, scaffold erected next to a wall enables a bricklayer to extend the wall to roof height. It includes working platforms, ladders and guardrails, and is usually constructed from steel or aluminium alloy tubes that are clipped or coupled together.

There are two basic forms of scaffold: putlog and independent.

Putlog scaffolds consist of a single row of uprights or standards set away from the wall at a distance that accommodates the width of the working platform. The standards are joined together with horizontal members called '**ledgers**' and are tied to the building with cross members called '**putlogs**'.

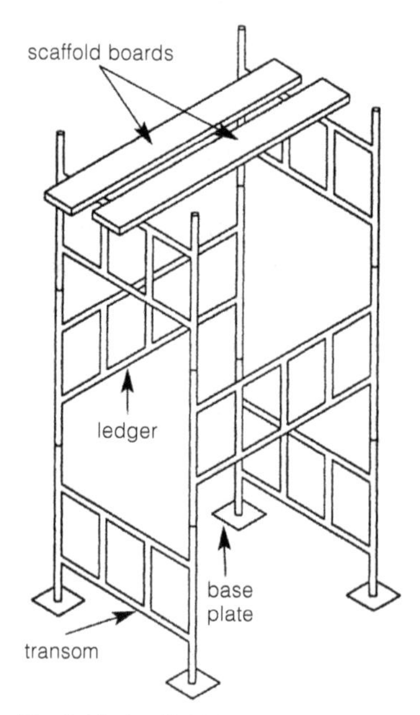

Fig 2.28 Scaffold frame

Independent scaffolds have two rows of standards that are tied by cross members called transoms. This form of scaffold does not rely on the building for support and is therefore suitable for most types of structures. For safety reasons, it may be necessary to tie the scaffold to the building at regular intervals using a horizontal tube called a 'bridle bearing'. Alternatively, reveal pins can be used to secure cross members and brace the scaffold.

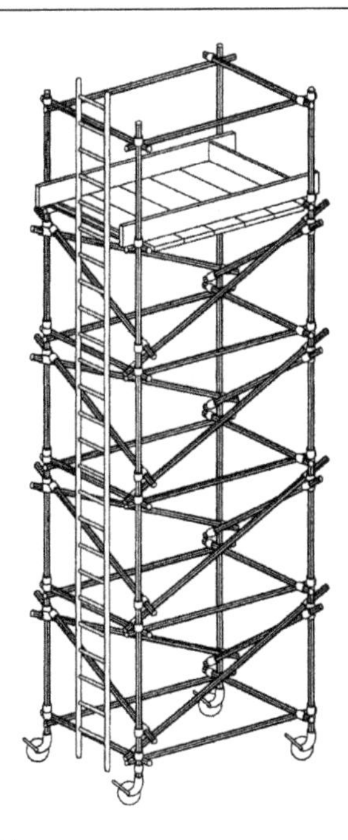

Fig 2.29 Putlog scaffold

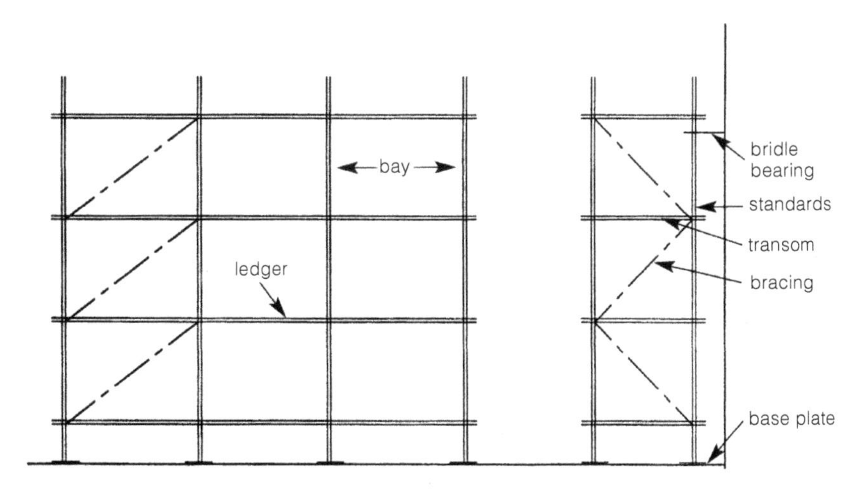

Fig 2.30 Independent scaffold

Scaffolding can be made from tubular steel, tubular aluminium alloy or timber.

Tubular steel can be welded or seamless tubes of 48 mm outside diameter. They can be either galvanised (to guard against corrosion) or ungalvanised, but ungalvanised tubes will require special care like varnishing, painting or dipping in an oil bath after use.

Aluminium alloy tubes are seamless tubes of 48 mm outside diameter. They do not need protective treatment unless they come into contact with damp lime, wet cement or seawater that can cause corrosion.

Using **timber** as a form of scaffold is seldom seen in South Africa, although it may still be done elsewhere. Instead of using coupling fittings (as with metal scaffolding), timber uses wire, rope, nails or bolts to tie the members together.

Scaffold boards are usually boards of softwood timber used to form the working platform. The dimensions are 225 mm × 38 mm and should NOT exceed 4.80 m in length. To prevent the ends from splitting, they should be bound with at least 25 mm-wide hoop iron at the ends. You can see where scaffold boards are normally placed in Fig. 2.28.

2.6 Formwork

Formwork for *in situ* concrete work may be described as a mould or box into which wet concrete can be poured to form the shape of the required member.

2.6.1 Materials

Timber is the most common material used for general formwork. It can be smooth or rough depending on the type of surface texture required. It is essential that the timber's moisture content is between 15 and 20% so that the moisture movement of the formwork is minimal. If the timber is too dry, it will absorb moisture from the wet concrete, which could weaken the concrete member. Swelling and bulging of the formwork could also occur, resulting in unwanted or unsightly shapes. If the moisture content of the timber used is too high, it may shrink, resulting in open joints and grout leakage.

- **Plywood** is extensively used to construct formwork units as it is strong and light. It is supplied in sheets measuring 1.20 m wide with standard lengths of 2.40 m, 2.70 m and 3.00 m. The thickness selected must relate to the anticipated pressures to which the plywood will be exposed.
- **Clipboard** can also be used for formwork but, because of its lower strength, it will need more supports and stiffeners. Also, clipboard cannot be used as many times as timber, plywood or steel.
- **Steel forms** are generally based upon a manufacturer's patent system. Steel is not as adaptable as timber but, if it is treated with care, it can be re-used 30 to 40 times (double that of timber).

Formwork must fulfil the following functions:

- It should be **strong** enough to support the load of wet concrete.

- It must **not deflect** under load, which includes the loading of the wet concrete, self-weight and any other superimposed loads such as construction workers, wheelbarrows, etc.
- It must be **accurately set out**. Concrete, when placed, is fluid and will take up the shape of the box or mould into which it is placed. The shape of the formwork must be correct, of the right size and in the right position.
- The joints must be properly **sealed** to prevent unnecessary loss of concrete or unsightly ridges and honeycombing of the finished concrete. Joints are normally sealed with flexible polyurethane strips or special adhesive tape.
- Formwork must be designed to be **manageable** in terms of handling and fixing in place.
- The materials used for formwork must be **easily fixed** using nails or wood screws. Nails should be a minimum of two-and-a-half times longer than the wood thickness.

Two defects can occur on the surface of finished concrete:

- **Uneven colour**. The irregular absorption of water from the wet concrete can lead to discolouring. When new formwork is used together with older ones, this can aggravate the problem.
- **Blowholes.** These are small holes of less than 15 mm diameter caused by air trapped between the formwork and the concrete face.

It is common practice to apply, by brushing or spraying, a release agent onto the inside of the formwork to alleviate or reduce these effects. One must be careful not to apply the release agent to the reinforcement as this can lead to loss of bonding between the concrete and the reinforcement.

How can these defects be overcome?

2.6.2 Foundation formwork

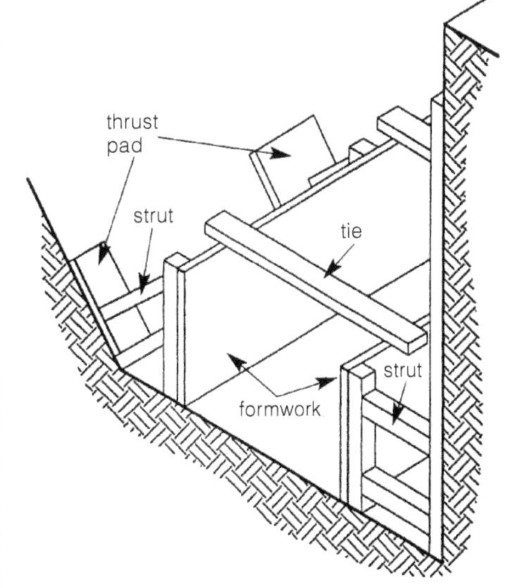

Fig 2.31 Typical foundation formwork

As the name suggests, this formwork is used in foundations like pad or isolated bases where subsoil conditions are unstable. If the subsoil is firm and hard, it may be possible to excavate the trench for the foundations to the size and depth required and cast the concrete against the excavated faces. Side and end panels of the formwork must be strutted against the excavation faces to resist the horizontal pressures of the wet concrete, causing irregular shapes or collapse.

2.6.3 Column formwork

A column form or box consists of a vertical mould that has to resist considerable horizontal pressure in the early stages of casting while the concrete is still wet. While casting the foundation, construction workers usually make provision for a 75 mm protrusion above the base called a **kicker**. It is against this kicker that the formwork for a column is located, to both accurately position the formwork and prevent the loss of grout from the base edge of the form. It is common practice to cast a column to its full height or up to the underside of the lowest beam. The panels forming the column sides can be strengthened using horizontal cleats or vertical studs (sometimes called 'soldiers'). It is important to keep the column formwork vertical. Stays are needed to keep the column box aligned.

2.6.4 Beam formwork

A beam form consists of a three-sided box supported by cross members. Column formwork often supports beam formwork. The **soffit** (underside of the beam) board should be thicker than the beam sides since this member will carry the dead load until the beam has gained sufficient strength to be self-supporting. When a slab needs to be cast, the formwork for the slab is often erected at the same time as that of the beam. The added advantage is that only one concrete pour takes place, minimising both time and effort.

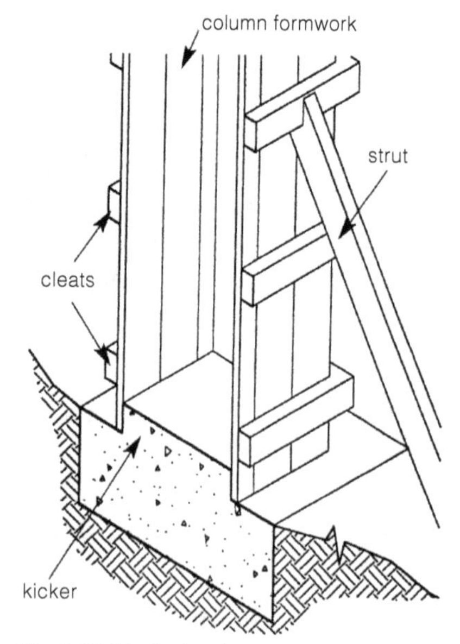

Fig 2.32 Typical column formwork

2.6.5 Slab formwork

All floor or slab formwork can be supported by beam formwork. It is essential to ensure that the slab formwork is level and this can be accomplished by adjusting the supports.

Once the formwork is erected, the following must be ensured:

- The insides of the forms are clean and clear of rubble or dirt before applying the releasing agent.
- All holes and joints are sealed to prevent grout loss.
- Spacer blocks are fixed to the reinforcement.
- The reinforcement is properly placed and fixed, and cannot be displaced during the pour.
- The concrete can be easily poured into the forms.
- Proper curing methods are applied.

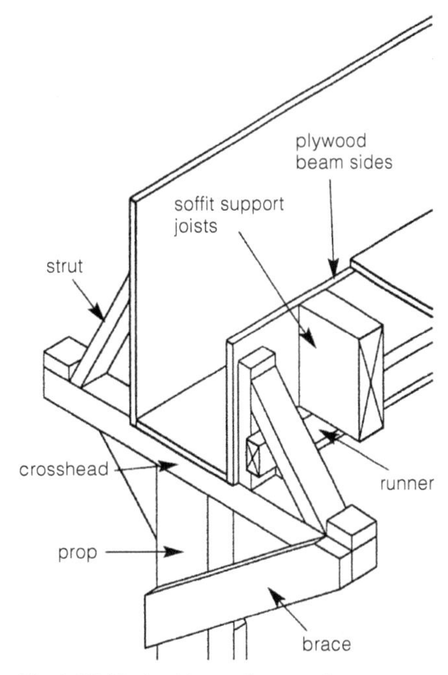

Fig 2.33 Typical beam formwork

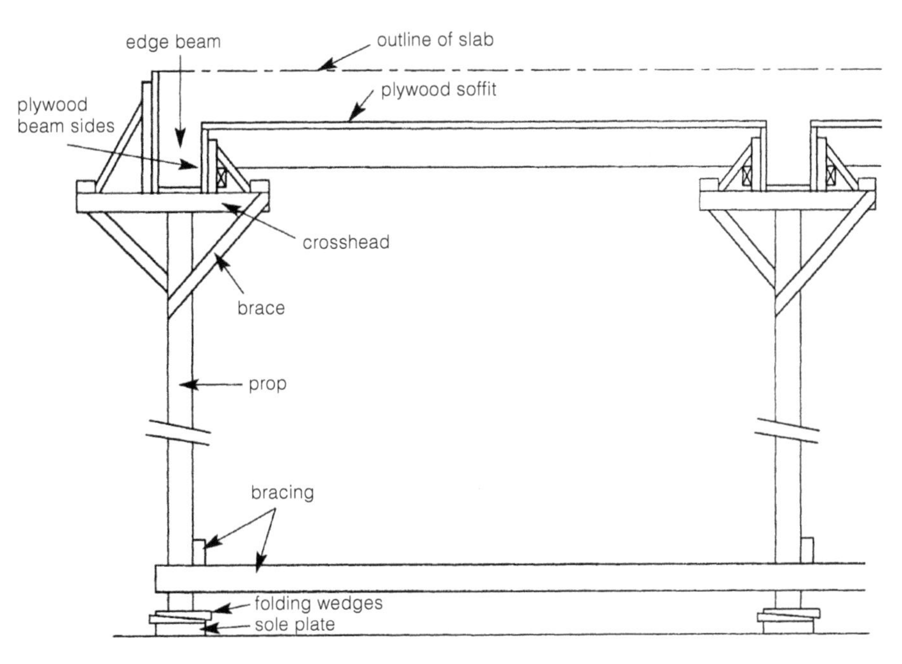

Fig 2.34 Typical slab and beam formwork

Do you remember the sections on reinforcement and curing in Chapter 2 of *Construction Materials*?

The formwork must be removed once the concrete has set sufficiently and when instructed to do so by the engineer. Once the formwork is removed, it should be carefully cleaned to remove any concrete that may have stuck to the face of the mould.

2.7 Concrete placing and compaction

When concrete is mixed on site (concrete mixer) or off site (ready-mixed concrete at a plant), one rarely finds that the mixture is discharged directly into the formwork. Ready-mixed concrete is fully mixed off site, and it is therefore important that the transfer of concrete from the mixing stage to where it is placed does not have any effect on the concrete's quality. In Chapter 2 of *Construction Materials*, we discussed the problems that could arise from using inferior quality concrete.

During the transfer and placing process, the following could occur:

- segregation (see Chapter 2, *Construction Materials*);
- contamination of concrete (dirt, oil, etc.);
- premature stiffening.

Have you noticed what method is used to transport concrete at a construction site?

2.7.1 Transport and placing

On a construction site, various methods can be used to transport concrete, depending on factors such as the accessibility of the site, the amount of concrete to be poured and the location of the pour.

You may have seen the following:

- **Wheelbarrow.** This is the most common means of moving concrete from the mixer to the pour. However, it is not suitable for ready-mixed concrete. Scaffold boards are often placed to provide access, but generally straight and nearly level areas are suitable.

Fig 2.35 Concrete dumper

- **Dumper.** This is like a mechanised wheelbarrow and able to carry larger quantities of concrete.

- **Crane and skip.** Where it becomes impossible to use wheelbarrows or dumpers, for example very tall buildings, contractors often use a crane fitted with a special bucket (skip) to transport concrete.

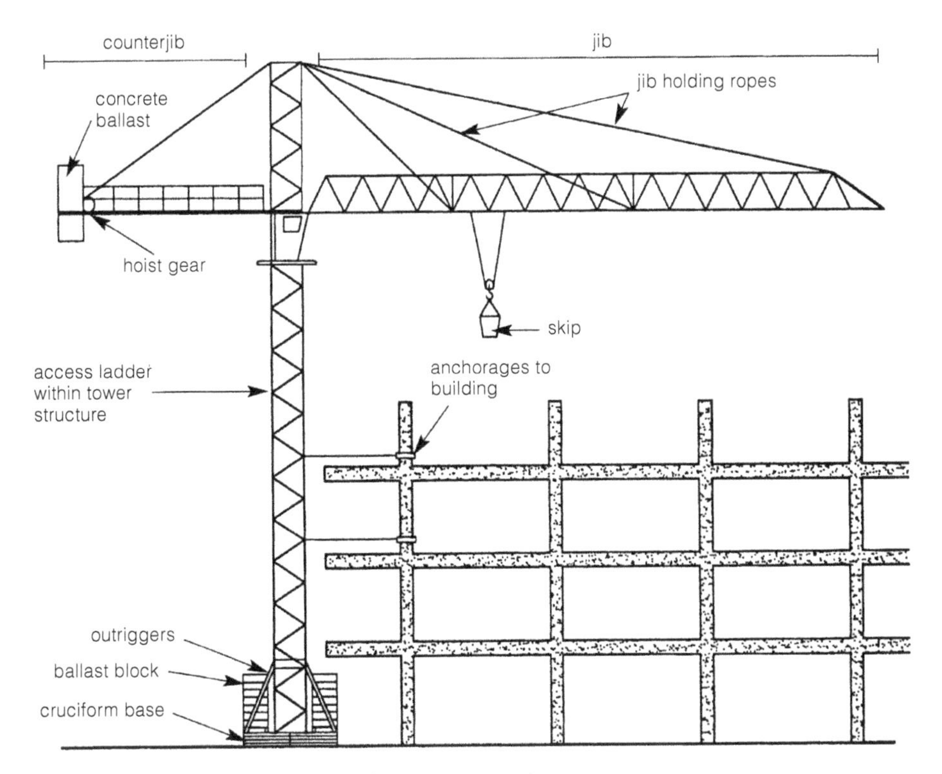

Fig 2.36 Crane skip operation

- **Ready-mixed concrete.** Many concrete trucks have half-round chutes (and extensions) to allow concrete to be poured directly from the truck into the formwork. However, the truck must be able to park close to the area where the pour is to take place.

Can you think of any other ways of transporting concrete? Is it necessary to compact concrete? How is concrete compacted? Do we take a vibrating roller and compact it for a number of passes, like in road construction?

The strength of concrete is not only the direct result of its mix proportions, but also the process of placing and compaction once it is in position. Quality plays an important role in the performance of concrete and it is essential to ensure proper work procedures (skilled workers, correct equipment and proper supervision).

When placing concrete, the following rules should be observed:

- Concrete should not be dumped onto the formwork. This could result in either the steel bending excessively or moving from its original position.
- Try to place the concrete as close as possible to its final position.
- Do NOT use the vibrator to move concrete excessively.
- Pour concrete at a steady rate.
- When working on inclined surfaces, start at the bottom and work upwards.
- Concrete should NOT be dropped from a height exceeding 3 m as this will result in segregation.

2.7.2 Compaction

Concrete needs to be compacted to drive out trapped (unwanted) air in the concrete mix. This results in:

- more durable concrete;
- smoother appearance;
- better impermeability;
- higher compressive strength; and
- a decrease in honeycombing (see Chapter 2, *Construction Materials*).

There are two ways of compacting concrete: manual (hand) or mechanical compaction.

Hand compaction

In slabs, you will often find that workers use a long scaffold board or a length of timber board to **tamp** the concrete. **Tamping** is done by dropping the board in short chopping motions onto the slab, thereby providing compaction. Have you seen this being done?

Rodding is another method of manual compaction that uses wooded rods. These rods are used in an action similar to when making concrete cubes or grinding corn in a pot.

Mechanical vibration

There are three ways of applying mechanical vibration:

- internal (inside the concrete);
- external (onto the formwork); and
- surface (on the surface of the concrete).

Internal vibration

A device called a **poker vibrator** is used to perform this function. They come in various sizes ranging in diameter from 25 to 150 mm (60 or 75 mm are more popular). Poker vibrators are powered by petrol/diesel motors, electricity or compressed air.

The poker should be inserted into the concrete quickly (at points approximately 300 mm apart) and removed slowly to allow the trapped air to escape.

Try not to allow the poker to touch the reinforcement or the formwork. The poker should not be used to move concrete around. Do not leave the poker running for long periods if it is not inserted into the concrete as this may result in damage.

How do you know when to stop compacting?

Compaction is complete when:

- no more **bubbles** appear on the surface of the concrete;
- the surface has a **glossy appearance**; and
- the **sound** of the vibrations changes.

External vibration

This method is often used in the precast industry where some form of mechanical vibration is applied directly to the formwork or via a vibrating table. The formwork must be strong enough to withstand this externally applied pressure.

Surface vibration

Surface vibration mainly applies to slabs where a vibrating roller or beam is placed in direct contact with the concrete surface. It is similar to tamping but involves a power-driven unit.

2.8 Brickwork bonding

When building with bricks, it is necessary to lay the bricks according to a **recognised pattern** or **bond** to ensure the stability of the structure and to produce a pleasing appearance. All bonds are designed so that no vertical

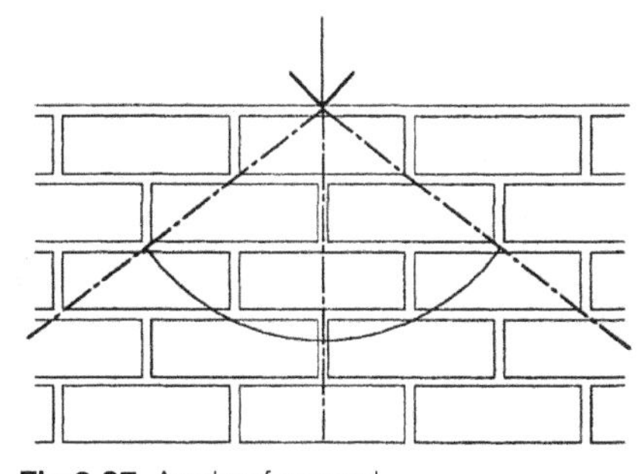

Fig 2.37 Angle of spread

joint in any one course is above or below a vertical joint in the adjoining course. To simplify this requirement, special bricks are produced or cut from whole bricks on site. The bonds are also planned to give the greatest practical amount of lap to the bricks (not less than a quarter of a brick length). Properly bonded brickwork distributes the load over as large an area a possible, with the angle of spread of the load being 60°.

2.8.1 Common bonds

The construction materials handbook highlighted the most common types of bonds used in brickwork. Here's a brief summary:

- **Stretcher bond** comprises bricks placed with their longest side facing the outside of the wall. This is used for half brick walls and cavity walls.
- **English bond** is a very strong bond consisting of alternate courses of headers and stretchers.
- **Flemish bond** consists of alternate headers and stretchers in each course. Its appearance is considered better than that of English bond, but is not quite as strong. This bond requires fewer facing bricks than English bond and is sometimes referred to as the double Flemish bond.

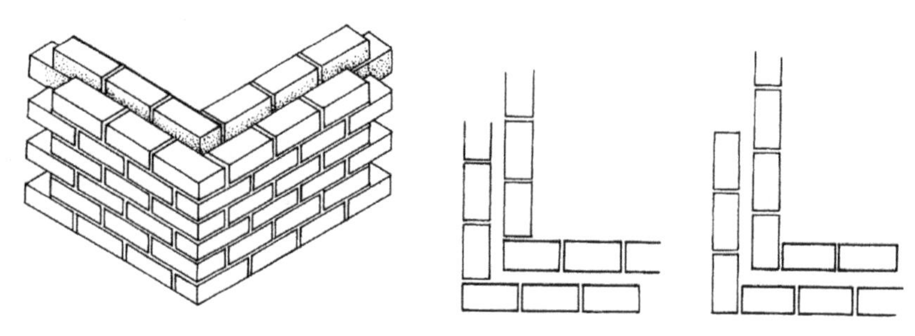

Fig 2.38 Stretcher bond

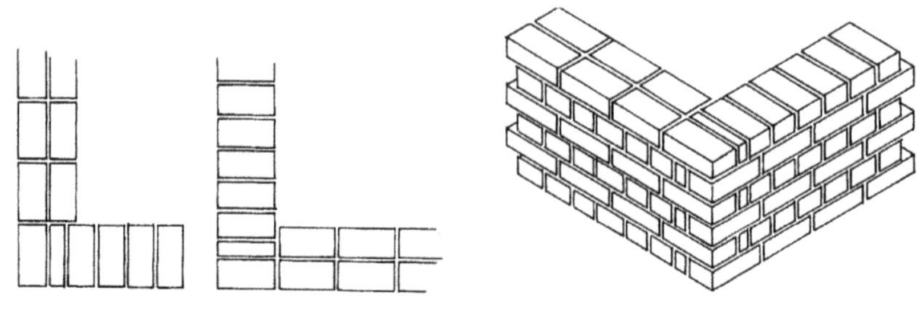

Fig 2.39 English bond

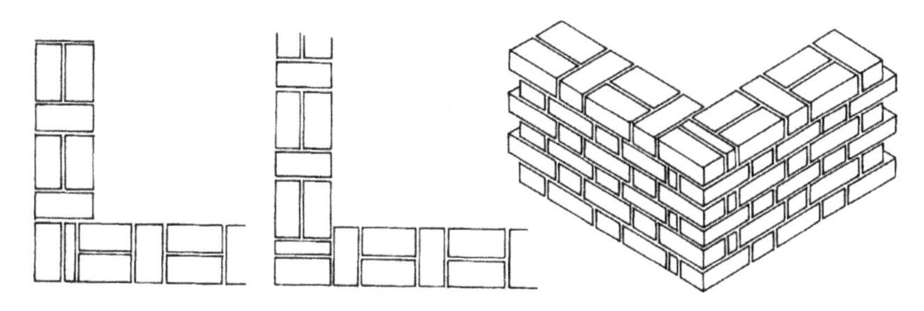

Fig 2.40 Flemish bond

2.9 Summary

The purpose of this chapter was to:

- Explain the term structures as a generic term referring to all forms of building work.
- Review the components of a structure either as concrete or steel.
- Explain the purpose of providing formwork to structural components.
- Identify and explain the importance of formwork to concrete.
- Identify and explain the importance of scaffolding in general construction.
- Identify the construction methods in the application of bricks.

Self-evaluation 2

1. Complete the sentences:
   a. The most important function of a building is to provide _____.
   b. A _______ is a vertical member that carries the beam and floor loads to the foundation.
   c. _______ structures consist of a series of beams and columns that are connected together to form an arch structure.
   d. Universal beam sections are described by their _______ and _______.
   e. _______ are used at the edges and ends of members.
   f. Problems associated with the use of black bolts relate to ___.
   g. _______ is a temporary structure from which people can gain access to high-level working areas.
   h. _______ is described as a mould or box into which wet concrete is poured to form the shape of the required member.
   i. Formwork must be designed to be _______ in terms of handling and fixing in place.
   j. When casting the foundation, the 75 mm protrusion above the base is called a _______.
   k. A device called a _______ is used to provide internal vibration for concrete.
   l. _______ comprises bricks placed with their longest side facing the outside of the wall.

➤

2. State whether the following are true or false:
   a. The construction process is concerned with the nature and sequence of operations on site.
   b. Y16 refers to the number of reinforcing bars required.
   c. Horizontal shear stresses result from the tendency of a beam to bend under load and split into horizontal laminations.
   d. Rivets are the cheapest bolts available.
   e. Beam to column connections can be designed as simple connections.
   f. Precast concrete is a mixture of cement, aggregate and water placed in a mould.
   g. Putlog scaffold does not rely on the building for support.
   h. It is not necessary to seal the joints when erecting formwork.
   i. Concrete should not be dumped onto formwork.
   j. Concrete needs to be compacted to drive out unwanted air in the concrete mix.
   k. The correct bonding ensures stability of a structure.
3. Answer the following short questions:
   a. What is the function of a structural engineer?
   b. What loads does an engineer consider when designing a structural member?
   c. Explain what you understand by the following notation: 12Y16-02-180B2.
   d. When using structural steel angles, describe your understanding of the following: 48 × 48 × 3.
   e. What is the difference between putlog scaffold and independent scaffolds?
   f. Describe the functions formwork must fulfil.
   g. Why is it so important to ensure that proper work procedures are followed when placing and compacting concrete?

Answers to self-evaluation 2

1. a. shelter
   b. column
   c. portal frame
   d. size and mass
   e. fillet welds
   f. shearing
   g. scaffolding
   h. formwork
   i. manageable
   j. kicker
   k. poker vibrator
   l. stretcher bond
2. a. true
   b. false
   c. true

d. false
e. true
f. true
g. false
h. false
i. true
j. true
k. true

3. a. see ref 2.1, page 36
   b. see ref 2.2.1, page 38
   c. see ref 2.2.1, page 38
   d. see ref 2.3.4, page 52
   e. see ref 2.5, page 58
   f. see ref 2.6.1, page 60
   g. see ref 2.7.1, page 64 and see ref 2.7.2, page 66

Chapter 3

Road engineering

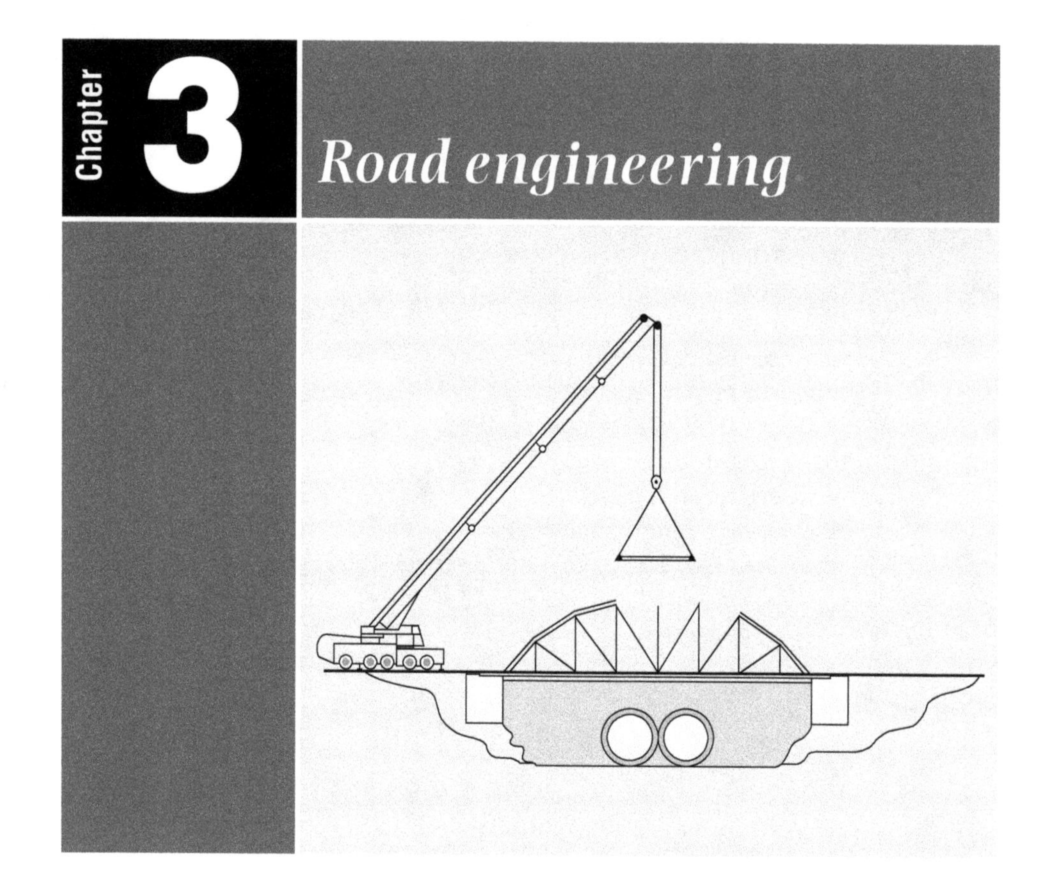

Outcomes

After studying this unit, you should be able to:

- Give the various modes of transportation
- Explain the process necessary when designing a road
- Discuss mass haul diagrams
- Identify the elements in the geometric design of roads
- Identify and understand the difference between flexible and rigid pavements and their properties
- Know and understand the different applications used as surface seals of the road
- Describe the importance of road drainage.

3.1 Introduction

Transportation is the movement of people and objects from one point (the **origin**) to another (the **destination**). In engineering terms this is commonly referred to as **O-D information** or origin–destination information.

The most common movement is the home-to-home work cycle where people need to get to and from work. This movement can take various forms and the study of these forms is called transportation engineering. Transportation is also considered to be a means of communication.

The most common means of travel are:

- pedestrian transport (walking);
- road transport (cars, buses, taxis, etc.)
- rail transport (trains);
- air transport; and
- water transport.

Fig 3.1 Origin–destination movement

Can you think of any other means of transport? How do you get to work, the shopping centre, classes, cinemas ...?

For the purpose of this course, we will concentrate on the general aspects of road construction in South Africa: earthworks, pavement design, pavement construction and drainage. Road construction follows basically the same process as any other construction project and involves:

- investigation;
- route location;
- road design;
- setting out work boundaries;
- construction; and
- opening of the road to the public for daily use.

3.1.1 Site investigation

Before any project is undertaken, there must be a need for that particular service. The need will be defined by any one of a number of role-players: national government, local authority, the public or any other parties that may have some interest in the project. Once the need has been identified, someone (the **client**) takes ownership of the project and, in most cases, makes funds available for the design and construction works.

Before the final location of road is fixed, several interim studies are undertaken:

- A desk(top) study;
- A preliminary investigation; and
- A detailed investigation.

Desk(top) study

This is the first step during which the beginning and end points of the road are fixed (i.e. the origin 'A' to the destination 'B'). Once these points have been fixed, a broad region will be determined in which all possible routes between 'A' and 'B' will be explored. This is done by establishing several broad bands (8–16 km wide in the case of major rural roads) in which to concentrate. To refine the search even further, several corridors (1–1.5 km wide) will be investigated.

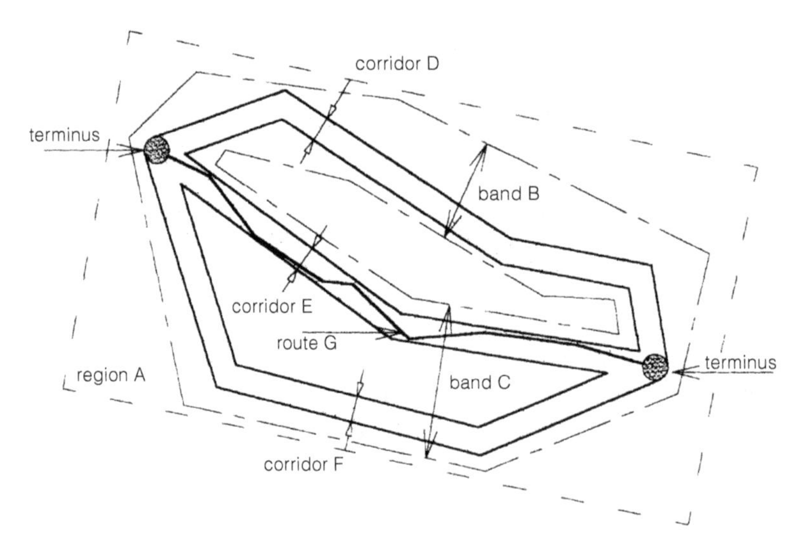

Fig 3.2 Desk(top) study

The best and most suitable corridor will be established, after which the alignment of the road will be determined. This takes place in the office as a desk(top) study before the actual design of the road takes place.

Preliminary study

The preliminary study is a large-scale investigation of the alignments established in the desk(top) study. At this stage, all the physical information that may affect the location of the road is collected. This includes:

- the position and invert levels of streams;
- the shape of the ground; and
- the position of trees, culverts, bridges, existing roads, etc.

The information may be acquired through aerial photographs or ground surveys (tacheometric and visual surveys). The information is then transferred onto maps and plans that will assist the engineer in designing the final alignment.

You are reminded that the above procedure is applicable to roads in a rural environment and the procedure is somewhat different in an urban environment. In an urban environment, there are strict town and urban planning policies in place that determine the form and layout that residential areas will take. Once the town planners have established the layout which will incorporate corridors for road alignments, the engineers will design accordingly. These corridors are referred to as **road reserves** within which the roads are designed.

Activity 3.1

Speak to your local human settlements practitioners (e.g. civil engineering consultants) and enquire from them how they go about designing roads in an urban environment.

You may find that some of the studies explained above are not entirely necessary, rather the focus will change to concentrate on what is prescribed by the planning layouts.

Detailed investigation

This survey serves the dual purpose of definitely fixing the centreline of the road and preparing final design plans for construction purposes. Detailed design will include horizontal and vertical alignments (discussed later in this unit). The culmination of the design will be realised by the pegging of the centreline and determining levels on site before construction begins.

With funding being drastically reduced from central government for road construction and maintenance, private initiatives have emerged. For example, some farmers in KwaZulu-Natal fund repairs to roadworks in order to improve the road network from their farms to nearby towns.

After funding has been secured, a competent person (or persons – usually a consulting engineering practice) is appointed to carry out the planning and design of the road.

Before the design of the road can begin, a thorough investigation of the prevailing conditions needs to be made. This investigation is carried out by means of aerial photographs and a site investigation, during which all the factors that have a bearing on the road's final location are considered.

When a road is built in an existing residential area, other services like drainage pipes, electrical cables, water pipes and telephone cables will be present in the area. The position and depth of these services must be established before excavation begins.

3.1.2 Route location

Once the route has been determined on the drawing, a field survey must be carried out to fix the centreline and determine ground levels along the proposed route. Final levels must be accurate as these form the basis for the next phase – the actual design of the road.

3.1.3 Road design

When a road is designed, several factors must be taken into account. The design can be compared to a recipe for baking a cake: if any ingredient is incorrect or missing, the cake will flop. The road design must cater for all the elements necessary to make travel safe. These include:

- cross sectional elements (e.g. width, cross fall, road reserve, lanes, etc.);
- design speed;
- sight distance;
- stopping/braking distance;
- vertical alignment;
- horizontal alignment;
- intersections; and
- pavement thickness.

The elements of design all relate to the road's geometry – in other words, what the completed road will eventually look like. There are several guideline documents used in civil engineering practice to determine and establish the factors to be considered when designing a road. These are available from organisations such as:

- CSIR – www.csir.co.za
- South African National Road Agency Limited (SANRAL) – www.nra.co.za
- Department of Transport (DoT) – www.transport.gov.za

Some of the publications like the *Technical Recommendations for Highways* (*TRH*) and *Urban Transport Guideline* (*UTG*) series are used extensively by road engineers.

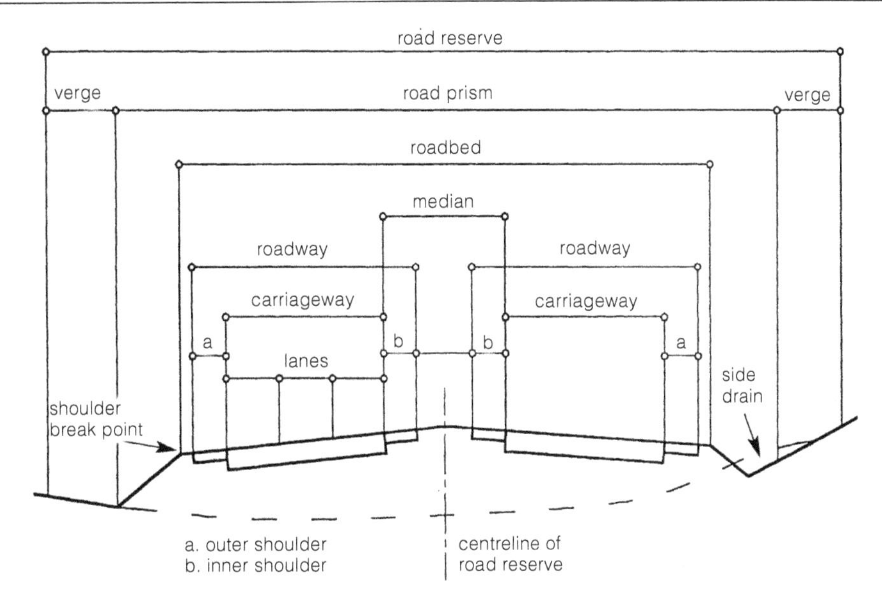

Fig 3.3 Road profile

3.1.4 Setting out

Once the road design is complete, a land surveyor will place and fix the final locations of the centreline and the left and right road edge positions with their associated finished road levels. This is the first step in preparing for the road's construction.

3.1.5 Construction

This is the longest and most intensive phase of the project cycle. This is when most of the activity takes place. A civil engineering contractor will be appointed (via tender) to construct the road. A site office will be established for the team to work from and construction plant will then be brought to the site.

What type of plant is necessary to construct a road?

The following are considered essential:

- bulldozers (for site clearing);
- compactors (for compacting the road layers);
- excavators (to install pipes beneath the road surface);
- road graders (to place and shape the road);
- trucks (to move material to and from the site); and
- water tankers.

Activity 3.2

Go to a road construction site and observe how it is done. Ask questions and then write a report about what you observed or share your experiences with other students.

Do you remember the work we did in Chapter 3 of *Construction Materials*, where we highlighted the structural layers of a road?

The road material will be placed in layers of varying thickness (usually 150 mm) depending on the designer's specification. Each layer must be compacted separately to achieve good founding and strength for the next layer. After compacting the layers, the surface will be covered with a wearing course that protects the layers beneath it. This wearing course can be a premix layer or a road seal combination of coarse aggregates and bitumen.

3.1.6 Opening the road to the public

Once construction is complete, the road is marked with lines and other road furniture, including traffic signals (robots), signage, etc. and the road is opened to the public. For national roads, a prominent person – for example, the Minister of Transport – is often invited to officially open the road.

Who is the current Minister of Transport?

3.2 Road terminology

- **Arterial highway** is a general term denoting a highway primarily used for through traffic, usually on a continuous route.
- **Freeway** is an expressway with full control of access, divided by an unbroken median.
- **Expressway** is a dual carriageway road with full or partial control of access.
- **Dual carriageway road** is a highway with roadways for traffic in opposite directions, separated by a median.
- **Service road** is a local street or road auxiliary to an arterial highway to maintain local traffic circulation and for control of access.
- **Roadbed** is the area extending from kerbline to kerbline or shoulder breakpoint to shoulder breakpoint.
- **Median** is the area between the two carriageways of a dual carriageway road.
- **Road reserve** is the area of land reserved for the construction and maintenance of the road, including the areas required for interchanges.

Why do we need roads?

South Africa is a developing nation with an annual increase in the number of vehicles being produced and purchased. To cater for this steady increase in vehicles and the demand for people to travel, one needs an extensive network of roads to allow safe travel at a suitable speed with minimum driver frustration. There are several agencies responsible for the network of roads within South Africa. This responsibility includes funding, design, construction and maintenance:

- National freeways, which used to be the National Department of Transport's responsibility, are now that of the National Roads Agency (NRA), also commonly known as the South African National Roads Agency (SANRAL), which obtains funds directly from central government.
- Major rural roads and arterials are managed by the provincial administrations and/or metropolitan councils, which receive funds from the allocation made to the SANRAL.
- Local streets are usually managed by the local authority; a municipality or local council.

3.2.1 Classification of roads

Roads in South Africa are classified according to their function and traffic characteristics. The classification system is as follows:

- **Class 1: National (primary) roads** – for example, the N1, N2, N3 – have limited access. In other words, you can only get onto a freeway at specified areas and these can be quite far apart. When next you drive on a freeway like the N1, see how easy it is to gain access to it from your home.
- **Class 2: Arterials** are high capacity roads that supplement the movement of traffic around a city. They give direct access to the freeways, but not to residential or industrial plots.
- **Class 3: District distributors** carry high volumes of traffic and are the links between higher and lower classes of road.
- **Class 4: Local distributors** are roads that link neighbourhoods and are often used as bus routes.
- **Class 5: Access roads** are further subdivided into cul-de-sacs, loops, etc., and form the network of residential developments. They give direct access to people's homes.

This classification is called a hierarchy and is graded according to a numbering system – number 1 being the highest grading and number 5 the lowest. The geometric configuration (the number of lanes, widths, speed, curves, etc.) is related to this classification. The higher the classification, the more stringent its safety factors.

Each class of road should intersect with at least two classes above or below it in the hierarchy.

Activity 3.3

Think about the area where you live. Does the road network conform to a natural hierarchy where you progressively move from lower order roads to higher order roads when undertaking long journeys? Give examples of each of the five classifications in identifying the network in your area.

3.3 Earthworks

Earthworks include removing topsoil, grading the exposed surface to the required formation level, preparing the subgrade and forming any embankments and/or cuttings.

The amount of earth that has to be moved will play a vital role in the final design of the road as hauling materials (transporting existing or imported materials) is an expensive operation.

Ground water and particularly the water table play an important part in the design of any road. We will look at the control of water and its influence on the road structure later in this chapter. It is, however, important to mention that water influences the strength of the subgrade. The moisture content of the subgrade must be optimal (not too high or too low) for greatest strength.

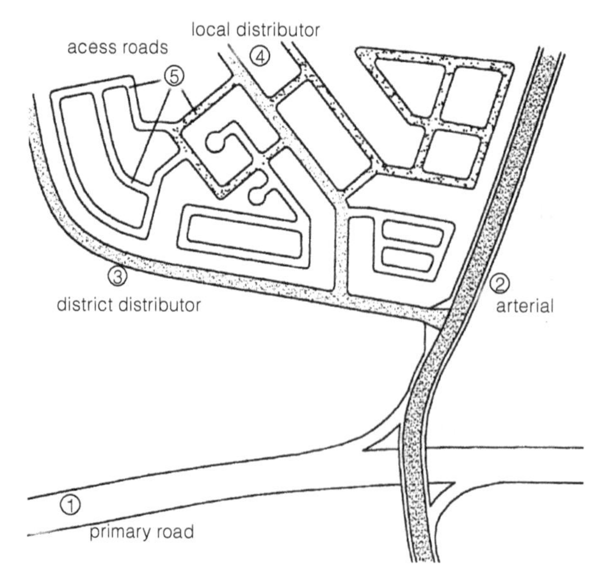

Fig 3.4 Road hierarchy

Why is it so important to have a fairly strong subgrade?

The subgrade fulfils the same function as a foundation in a building. If the foundation is weak, the building will eventually crack as it settles. The greater the settlement, the larger the cracks will be until possible failure occurs. Similarly, a road needs a strong foundation, mainly because of the nature of the loads applied to it. To an extent, the layers of soil above the subgrade will try to dissipate and spread the load, but the subgrade will still have to provide some support. Without this support, the road will fail and signs of this will be cracking, potholes, etc.

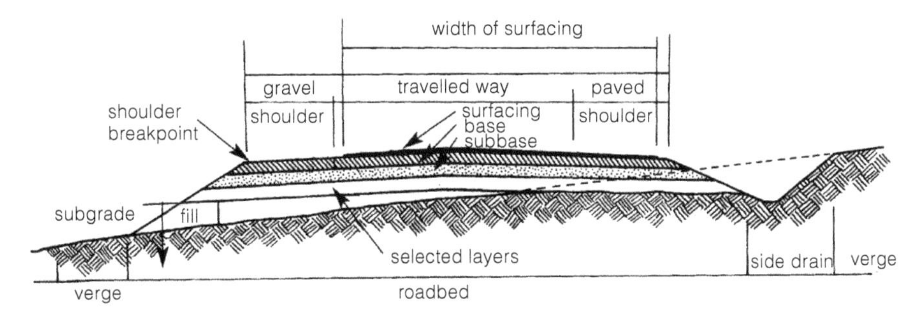

Fig 3.5 Road cross-section

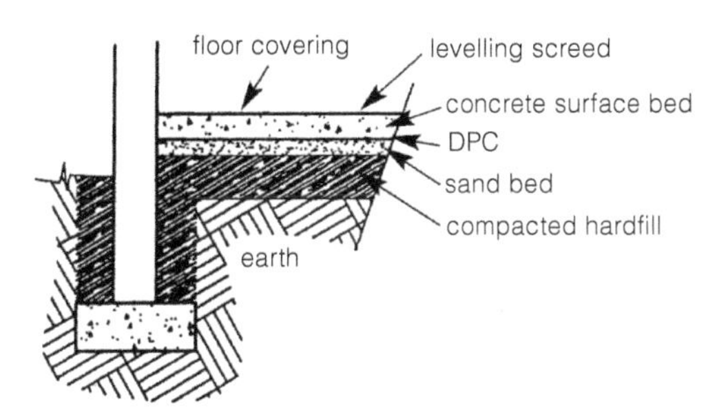

Fig 3.6 Typical house foundation

The structural layers above the subgrade will protect it against moisture ingress from above. However, during the construction period, the subgrade should be protected from the ingress of water.

Depending on the topography of the area through which the road passes and the final formation level of the road, the excavation of excess material or filling of imported material will need to take place. Two terms are synonymous with earthworks: **cut** and **fill**.

We know that the natural profile of the ground is not always level and one often encounters hills or valleys, particularly in areas of KwaZulu-Natal. Most times it is not very economical to design a road that follows the natural profile because of factors like the effect of steep inclines on slower moving traffic. The road designer then tries to reach a compromise by balancing the amount of earthworks and this is done by altering the alignment and profile of the road. In section 3.4 of this chapter, you will briefly be introduced to the various components of road geometry – horizontal and vertical alignment. In constructing a road through a hill, the material excavated during this process can be used to fill some of the lower lying areas.

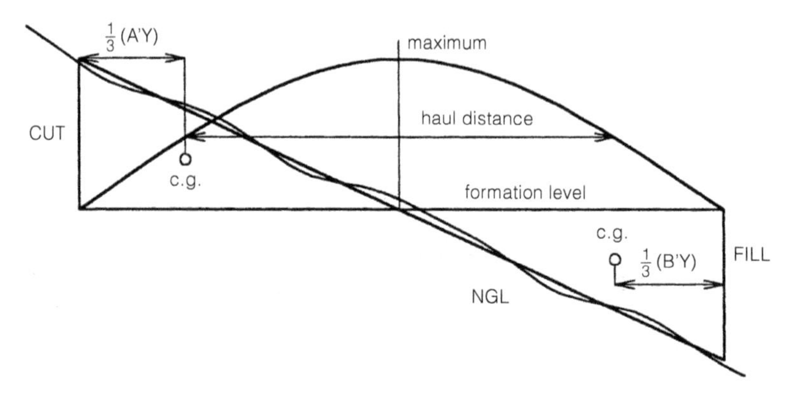

Fig 3.7 Balancing of earthworks

Cut is the removal of surplus material to reach the formation level of the road and fill is using this excess material or other material to raise the level of a low-lying area to the formation level. Fill areas are known as embankments. The fill material is spread in layers and compacted to avoid settlement at a later stage.

A cross-section is a graphical representation of a vertical section perpendicular to the length of road showing the ground and the proposed formation level of the road that is used for calculating earthworks. It will give the engineer or technician an idea of how much material will need to be cut or filled, in other words, disposed or imported material.

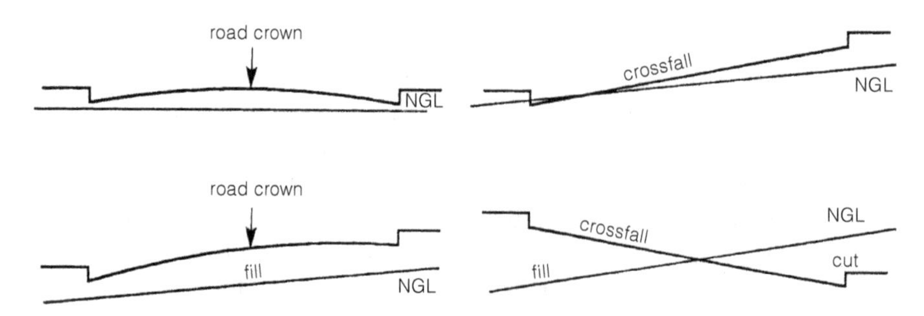

Fig 3.8 Road cross-sections

3.3.1 Definitions

- **Haul** refers to the volume of material multiplied by the distance over which it is moved.
- **Overhaul** and **freehaul**. A contractor may offer to haul material a distance of, say, 250 m at a cost of R1.00 per cubic metre. The contract may specify this amount. For any distance hauled beyond 250 m, the contractor may charge an extra 50c per cubic meter per 100 m of material moved (R0.50c/m^3/100 m). The distance of 250 m is called the freehaul distance and is based on the economical hauling distance of the construction plant used. The haul beyond the freehaul distance is called the overhaul.
- **Waste/spoil** is material excavated from the cuts but not used for embankment fills.
- **Borrow** is material needed for the formation of embankments, obtained from somewhere other than the roadway excavation (a borrow pit).
- **Station metre (Stn.m)** is 1 m^3 of material moved 100 m. In other words, 20 m^3 of material moved 1 500 m is a haul of (20 × 1 500)/100 = 300 Stn.m
- **Shrinkage/bulking**. Soil removed from its natural state is classed as loose volume and, when that same soil is needed to fill an embankment, a reduction in the volume of material (shrinkage) takes place, especially due to compaction. It must be understood that this factor only applies to material utilised from cut volumes and used for fill. The type of material will influence the factor but, generally, a factor of 0.85 is used. For example, 20 m^3 of material will be used for fill and therefore this volume must be multiplied by a factor of 0.85 (i.e. 20 × 0.85 = 17.5 m^3) allowing for compaction. Shrinkage is therefore the decrease in volume of earthwork material after placing and compaction. Bulking is an increase in the volume of earthwork material after excavation. A factor is applied in recognition of this phenomenon. Depending on the type of material, the factor can change but usually this is about 1.2. Should we therefore use a similar volume of material above, then the adjusted volume will be (20 × 1.2) = 24 m^3.
- The **datum** is a horizontal reference plane or level from which heights or depths are calculated.
- **Chainage** is the distance measured along the centreline that is used to identify the distance from the starting point, usually at intervals of 100 m.

3.3.2 Mass haul diagrams

To help quantify the amount of earthworks – cut or fill – and the distances the material needs to be moved, the road designer will draw a mass haul diagram (MHD) as a graphical representation of soil volumes.

A **mass haul diagram** is a graphical representation of the amount of earthworks involved in road projects. It shows accumulated earthwork volumes at any point along the proposed centreline of the road and the manner in which these volumes may be most economically handled.

Mass haul diagrams are used to compare the economies of various methods of earthwork distribution. Using the MHD, plotted directly below the longitudinal section of the surveyed centre line, one can work out the following:

- the distances over which cut and fill will balance;
- quantities of materials to be moved and the direction of movement;
- areas where earth may have to be borrowed or wasted (spoil) and the amount involved; and
- the best policy to adopt to obtain the most economic use of construction plant.

More about the MHD and its balancing procedures can be found in the subject Drawing II.

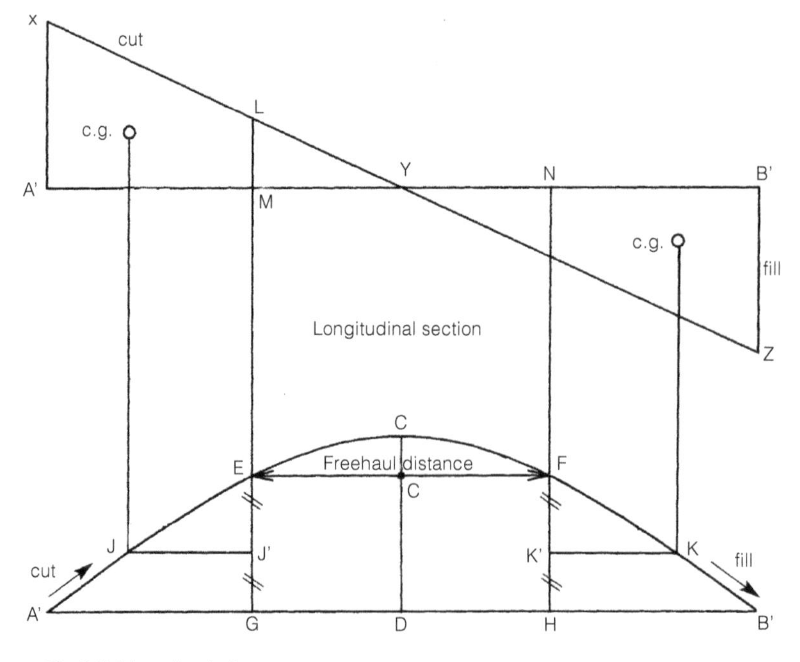

Fig 3.9 Mass haul diagram

Activity 3.4

The following notes refer to a 1 200 m section of a proposed road. The earthwork distribution in this section is to be planned disregarding the adjoining sections. The table shows the stations with the ground levels along the centre line. The formation level of the road is at an elevation above datum of 43.50 m at chainage 70 and rising uniformly on a gradient of 1.2%. The volumes are recorded in m^3; the cuts plus (+) and the fills minus (–).

Table 3.1 Cut and fill volumes

	Natural ground level	Volume	
		Cut (+)	Fill (–)
70	52.80	0	
71	57.30	1 860	
72	53.40	1 525	
73	47.10	547	
74	44.70		238
75	39.70		1 080
76	37.50		2 025
77	41.50		2 110
78	49.60		1 120
79	54.30		237
80	60.90	362	
81	62.10	724	
82	78.50	430	

1. Plot the longitudinal section using a horizontal scale of 1:1 000 and a vertical scale of, say 1:50. Use your drawing skills developed in the *Drawing* handbook to draw the longitudinal section, plotting the natural ground levels from the above table against the rising road profile, starting at level 43.50 m and calculating levels along the road line using the grade of 1.2%.
2. Assuming a correction (shrinkage) factor of 0.8 is applicable to fills, plot the MHD to a vertical scale of 1 000 m^3 to 20 mm.
3. Calculate the total haul in station metres and indicate the haul limits on the curve and section.
4. State which of the following estimates you would recommend:
   a. No freehaul but 50c per cubic metre for excavating, hauling and filling.
   b. A freehaul distance of 300 m at 40c per cubic metre and 20c per station metre for overhaul.
5. Apply the correction factor to the fill volumes as indicated in the table on the next page. ➤

Table 3.2 Corrected fill volumes

	Natural ground level	Volume		Corrected fill volumes (fill × 0.8)
		Cut (+)	Fill (−)	
70	52.80	0		
71	57.30	1 860		
72	53.40	1 525		
73	47.10	547		
74	44.70		238	(238 × 0.8) = 190
75	39.70		1 080	?
76	37.50		2 025	?
77	41.50		2 110	?
78	49.60		1 120	(1 120 × 0.8) = 896
79	54.30		237	190
80	60.90	362		
81	62.10	724		
82	78.50	430		

6. Now calculate the cumulative volumes in order to plot the MHD. The cumulative volumes are the algebraic values of the volumes in the cut (+) and corrected fill (–) columns.

Table 3.3 Cumulative volumes

	Natural ground level	Volume	Corrected fill volumes (fill × 0.8)		Cumulative volumes
		Cut (+)	Fill (−)		
70	52.80	0			0
71	57.30	1 860			+1 860
72	53.40	1 525			(+1 860 + 1 525) = +3 385
73	47.10	547			
74	44.70		238	190	(+3 932 – 190) = +3 742
75	39.70		1 080	864	
76	37.50		2 025	1 620	
77	41.50		2 110	1 688	(+1 298 – 1 688) = −430
78	49.60		1 120	896	
79	54.30		237	190	
80	60.90	362			
81	62.10	724			(−1 154 + 724) = −430
82	78.50	430			0

7. Once the MHD is plotted, determine freehaul and overhaul.

Read section 10.1 in the *Drawing* handbook for an example of how to complete the calculations and draw an MHD.

3.4 Geometric design of roads

Geometric design is the arrangement of the visible elements of a road, such as the horizontal and vertical alignments. There are several manuals and computer software packages that allow road designers to manipulate information necessary for the design of any road. Among the manuals used is the *TRH17, Geometric Design Manual of Provincial Roads Department.* For any project, an engineer or technician will have to prepare the following:

- the contours and grid system;
- the design and drawings of the horizontal alignment;
- the design and drawing of the vertical alignment;
- typical cross sections;
- cross-sections at regular intervals for calculating earthwork quantities; and
- design and drawing of intersections.

3.4.1 Contours

Plans showing contours are needed for many different types of civil engineering work, and in almost all cases where extensive earthwork excavations are involved. These plans are not just used for choosing the site or route of a road and planning the work in the most economical way, but also as a basis for estimating and the calculating of earthwork quantities. The following are normally aimed for when preparing the contours plan on original drawings:

1. A grid system, for example either at 50 m or 100 m intervals to a suitable scale of, for example. 1:200 or 1:1 000 where:
   a. the grid lines are drawn in blue;
   b. grid values are written in black – for example, 16 700 x and 49 100 y;
   c. x-values increase from the top of the page downwards (because we are in the southern hemisphere); and
   d. y-values increase from right to left.

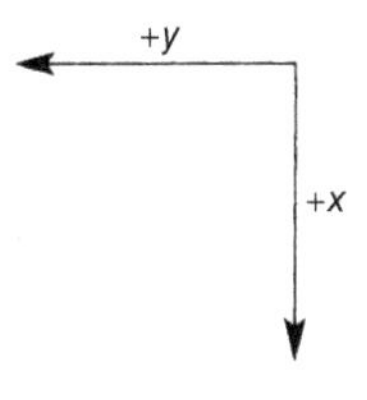

2. Contours are drawn at suitable intervals, for example 1.0 m, and the contour colour is burnt sienna (reddish brown). Contours at

5.0 intervals are darkened and written in bold to differentiate them from the 1.0 m contours.

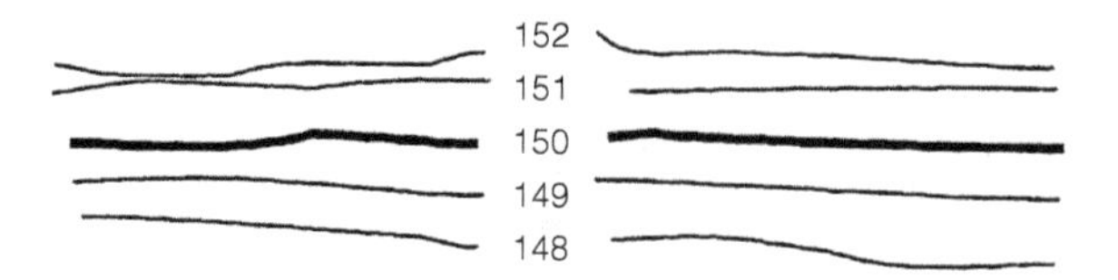

3. All existing structures and services are shown on the contour plan.
4. The scale is clearly indicated on the drawing.
5. The north sign is shown clearly in the direction of decreasing x axis but parallel to y axis.

3.4.2 Horizontal alignment

Horizontal alignment is the plan showing the route of the road and the different types of curves involved. A change of direction, whether left or right, in the horizontal plane relates to horizontal curves.

A horizontal alignment of a road is usually a series of straights (tangents) and circular curves connected by transition curves. Transition curves are curves of constantly changing radius that provide smooth movement from the straight to the circular curve. The point at which two straight lines (tangents) meet is called the point of intersection (PI). All curves will have a beginning of curve (BC) and an end of curve (EC) denoting where the curve starts and stops respectively. In other words, the BC and EC indicate the change from the straight to the curve or vice versa. These curves are designed with a minimum radius for a given design speed. For example, a road with a design speed of 70 km/h will have a minimum horizontal curve of radius 160 m. These values can be found in tables in the design manuals.

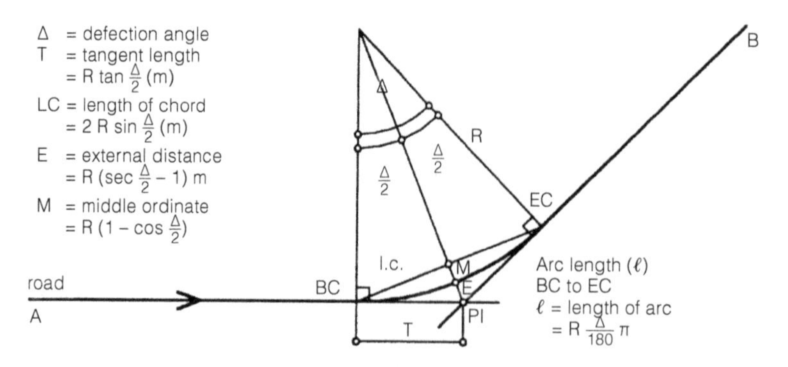

Fig 3.10 Circular curve

As a vehicle moves through a circular curve, centrifugal forces (pushing outwards from the centre) act on the vehicle. For a given radius and speed, a set force must be applied to maintain the vehicle in this circular path. When designing a road, this force is provided by the side friction developed between the tyre and the pavement. Should this be insufficient, the cross fall of the road is increased to counter the centrifugal forces caused by going around the curve. The increase in the cross fall is called super-elevation.

Super-elevation is the practise of sloping the carriageway of roads in horizontal curves to counteract the outward-acting forces on vehicles moving through these curves. It therefore relates to the difference or change in levels of the sides of the road/carriageway.

The maximum comfortable speed on a horizontal curve depends on the radius of the curve and the super-elevation. The desirable minimum radius of national roads is 1 500 m. The absolute minimum radius for 120 km/h is 540 m.

It is sometimes necessary to widen the road space needed when rear wheels follow a path of shorter radius than the front wheels. This is especially necessary for longer vehicles like buses and haulage vehicles. Have you noticed this when standing near a sharp bend in the road or watching a large vehicle turning? There is also a psychological demand for some clearance in the road space to divert a fast-moving vehicle safely around the curve.

3.4.3 Vertical alignment

The vertical alignment or longitudinal section is a section vertically through the centre line of a road structure to show the original and final ground levels.

We experience vertical alignment through the up and down movement we feel when we drive along a road. In other words, it is the change in level of the road viewed along the y-axis.

Between two straights of a road we insert a vertical curve, which is like a horizontal curve except that it takes place in the vertical plane. Transition curves may also be applied to vertical alignments, providing a smooth movement through the vertical plane. Curves are concave-shaped upwards (sag) at the foot of a valley or convex-shaped upwards (crest) at the summit. It illustrates a gradual change from one slope to the other.

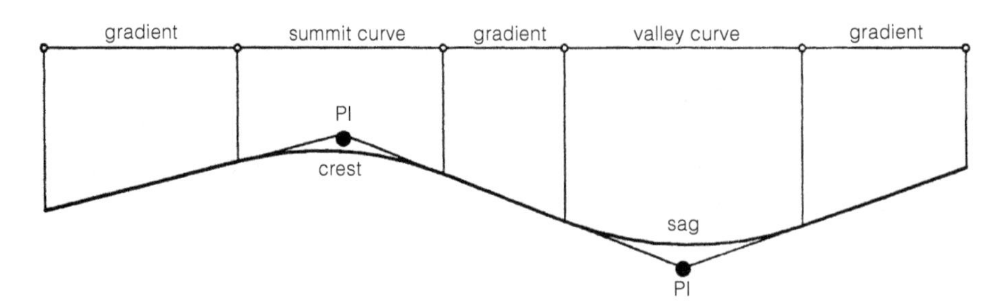

Fig 3.11 Sag and crest curves

Crest curves

These curves are controlled by sight distance. A comfort factor, commonly called a 'K-value', is obtained from tables in the design manuals. It is based on stopping sight distance and is needed to calculate the length of a vertical curve. The K-value for a crest curve at a design speed of 70 km/h is 23. This value is a factor and therefore does not have any units.

Sag curves

These curves are governed by the headlight illumination distance, especially while driving at night. Using the tables from the design manuals, a K-factor value of 20 is obtained with reference to a design speed of 70 km/h.

To expand your knowledge of the properties of vertical curves, look in the S2 *Drawing* handbook. The actual design and its elements will be studied in later courses.

To determine the length of both crest or sag curves, use the formula:

$$L = K \times A$$

where

L = length of vertical curve (m);

K = factor applied indicating the distance required to effect a unit change of grade and is affected by the design speed; and

A = algebraic differences (in percentages) between the grades on either side of the curves = G1 – G2

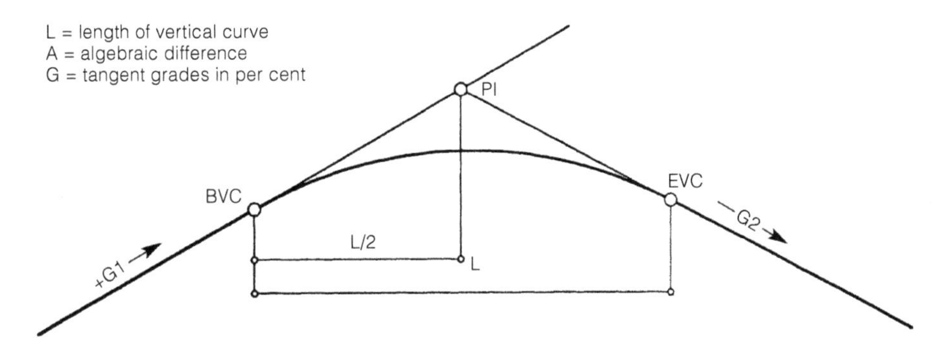

Fig 3.12 Vertical curve

Activity 3.5

How do we calculate the levels along a tangent?

Solution

We know that a tangent is a straight line starting at a known point (A) and ending at another known point (B). If the points are known, they should have two further criteria, i.e. a stake value and an elevation – for the tangent above, you will therefore have four knowns.

You can calculate the distance between the two points (A & B) by subtracting the two stake values from each other and similarly the elevation. Once the difference in elevation and the distance is known, you can determine the percentage gradient of the line.

To determine elevations at regular intervals along the line, you use the distance to the point to be calculated and multiply it by the percentage of the gradient and then add this value to the previous known elevation.

Let's say there is a tangent with two known points, i.e. A & B, with the following known information:

- Point A has a stake value of 1000 m and an elevation of 105.00 m.
- Point B has a stake value of 1200 m and an elevation of 110.00 m.

The difference in elevation is therefore 110.00 2 105.00 5 5.00 m.
The difference in distance is 1200 2 1000 5 200 m.
In order to calculate the percentage gradient along the tangent, we divide the elevation by the distance and multiply by 100, i.e. $\frac{5.00}{200} \times 100 = 2.5\%$.
Should we now be required to present elevations at every 20 metres, this can be easily calculated.

HINT: Draw a table to represent the information. 

Because we are determining levels at a regular interval (or offset), this means that the distance and gradient remains constant, i.e. 20 m and 2.5%. We can then calculate a constant factor to be used when adjusting the elevations at these regular intervals, for example:

$20 \text{ m} \times \frac{2.5}{100} = 0.500 \text{ m offsets.}$

What this means is that providing the distance of 20 m and the gradient of 2.5% remains constant, the offset will not change.

What does this mean – from Point A with an elevation of 1000 m we can add (because it is a positive gradient) a constant value of 0.500 m to determine the level at each 20 m chainage (stake value).

Complete the table:

Table 3.4 Calculating the levels along a tangent

Description	Distance (stake value)	Elevation
Point A	1000	105.00
	1020	105.50
	1040	106.00
	1060	106.50
	1080	107.00
	1100	
	1120	
	1140	
	1160	
	1180	
Point B	1200	110.00

EXAMPLE 3.1

Two tangents intersect at a point of intersection (PI). A vertical crest curve is required to connect these two tangents. Tangent 1 has a gradient of +1.13% while tangent 2 has a gradient of –0.82%. Design the vertical curve if the design speed of the road is 100 km/h. (Hint: Use the table to find the K-factor.)

Table 3.5 Minimum values of K for vertical curves

Design speed (km/h)	Crest curves	Sag curves
40	6	8
50	11	12
60	16	16
70	23	20
80	33	25
90	46	31
100	60	36
110	81	43
120	110	52
130	133	57
140	163	64

Solution

Using the table above, the K-factor for the crest curve can be obtained corresponding to the design speed of 100 km/h – i.e. **K = 60**

The algebraic difference in grades = [+1.13 – (–0.82)]
= 1.95

Therefore, using the formula L = KA
= 60 × 1.95
= 117 m

Theoretically, the minimum length for this curve is 117 m but, practically, it would be easier to round it off to 120 m as it makes it easier for setting out.

3.5 Pavement construction

Pavement is a general term for any paved surface and is also applied specifically to the entire construction of the road above the formation level. The pavement includes the subgrade, sub-base, base course and surfacing. A road pavement should not be confused with that hardened walking area next to a road that is referred to as a sidewalk.

Road pavements can be classified as **flexible pavements** that, for the purpose of design, are assumed to have low tensile strength and consist of a series of layers of materials to distribute the wheel loads to the subgrade. The alternative form is the **rigid pavement** of which, for the purpose of design, the tensile strength consists of a concrete slab resting on a granular base or sub-base.

There is another form of pavement construction which is gaining popularity and that is **block paving**. This pavement structure also uses layers in support of a top structure that replaces the asphalt with either cement or clay paving blocks.

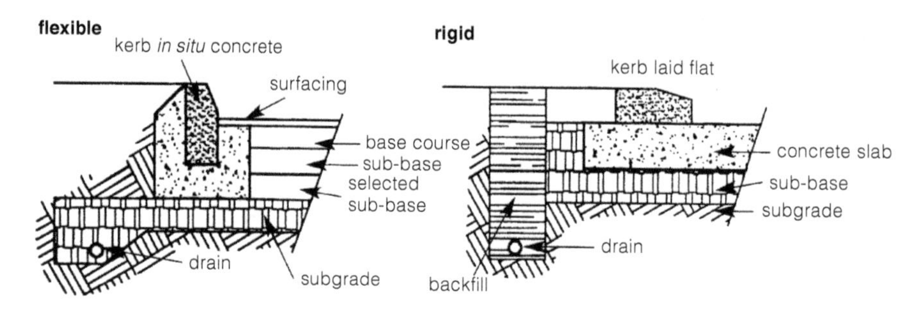

Fig 3.13 Pavement structures

3.5.1 Subgrade

The subgrade forms the formation level – this generally consists of the natural *in situ* material. In some cases, where the subgrade strength is poor (unable to safely bear the tensile stresses of the wheel loads), it is necessary to import more suitable material of a higher strength. This material is referred to as **selected subgrade (SSG) material.** In cases where the subgrade is within certain strength parameters, the road contractor will need to **rip and recompact (R&R).** To rip and recompact involves loosening the existing formation (subgrade), breaking up the material, watering, mixing and compacting it to a specified density so that the final compacted thickness of the prepared subgrade is not less than 150 mm. Care must be taken that no grass, roots or boulders greater in diameter than half the depth of the layer (75 mm) or vegetable matter is present in the prepared subgrade. These can lead to:

- root regrowth that can create havoc;
- uneven compaction; and
- decaying and rotting vegetable matter creating hollow areas or areas of uneven compaction in the subgrade.

Many of the terms described on the next page appear in *Construction Materials*.

3.5.2 Sub-base

The sub-base for a flexible pavement is laid directly on the formation level or imported SSG and should consist of well-compacted granular material. Most sub-bases are commercially produced (at a quarry) based on a blended mix of fine and coarse aggregate to attain a certain strength and exhibit certain specific properties. The actual thickness of the sub-base required is determined by the cumulative number of standard axles to be carried by the pavement over its design life and the strength of the subgrade determined by the California bearing ratio (CBR) test.

A **standard axle** is taken as 8 200 kg or 80 kN. The cumulative damage effect of all individual axle loads is expressed as the number of equivalent 80 kN single-axle loads (E80s). This is the number of 80 kN single-axle loads that would cause the same damage to the pavement as the actual spectrum of axle loads. The CBR test is an empirical test that measures the strength of a material relative to its thickness. The lower the value, the weaker the material. CBR values for sub-bases range between 45 to 60, depending on the material used. Only approved soil or gravel may be used and must be obtained from approved borrow pits. The material used must comply with certain requirements and adhere to approved specifications – these material specifications can either be supplied by an engineer or guidelines can be found in *TRH 14*.

The material is dumped onto the road and spread by means of motor graders, over the width of the layer. After the material has been spread, it must be prepared, watered, mixed and compacted using vibratory rollers. The sub-base must be compacted to the specified density at optimum moisture content for the type of equipment used. For most flexible pavement designs, the sub-base is approximately 150 mm thick after compaction.

The roads engineer may also decide to strengthen the sub-base even further by applying cement to the material. This material will then be referred to as a cement-stabilised sub-base.

3.5.3 Base course

In South Africa, there is a variety of base course material used, depending on construction techniques, availability of material and performance. There are three main categories:

- natural material base course (crusher run, gravel);
- asphalt bases; and
- concrete bases.

Natural base courses

Gravel base course

Only approved gravel obtained from approved borrow-pits may be used. Coarse aggregate must have a dimension after compaction not exceeding two-thirds of the compacted layer thickness. If the layer is 150 mm thick, the coarse aggregate size should not be greater than 100 mm. Oversize material may be screened out before placing, broken down on the road or bladed off during placing.

The gravel is dumped on the road and spread using a motor grader. The material is then prepared, watered, mixed and compacted using the appropriate compaction plant. Compaction must be done at optimum moisture content and the thickness of the compacted layer must not exceed the specified design.

Crusher run base course

Crusher run is a graded mix of coarse and fine aggregate (including crusher dust) passed through nine sieve sizes ranging from 53 mm to 0.075 mm.

Asphalt base course

An asphalt base course is a combination of fine and coarse aggregate and bitumen that is heated and mixed at a plant. The hot asphalt is transported to the site and placed on the road using a road paver. As soon as the asphalt is placed, it must be compacted using rollers. These bases are commonly referred to as bitumen treated bases or BTBs.

Concrete bases

The concrete is either mass concrete or reinforced concrete. Usually the concrete base includes a surfacing, but in some cases a 'black top' may be laid over it. Unreinforced concrete bases are constructed in slabs not exceeding 5 m spans with joints between slabs. These joints can also be reinforced with steel bars referred to as 'dowels' for added strength. The dowels are placed in the concrete after casting but before the concrete has achieved full strength.

Reinforced concrete slabs have reinforcing bars cast in the slab during construction and can either be a continuously reinforced concrete slab or one which is jointed.

3.5.4 Prime and tack coats

The **prime coat** is a bituminous binder applied to a non-bituminous granular base to provide a bond between the base and the surfacing. It also provides an additional seal against the ingress of water and binds the upper layer of the base material.

The **tack coat** is also a bituminous binder sprayed onto an existing bituminous surface or asphalt base to ensure adhesion of the new surface to the old underlying surface.

Prime coats are generally low viscosity cut-back bitumen, whereas tack coats are bitumen emulsions.

3.5.5 Surfacing

Surfacing should fulfil the following functions:

- skid resistance in all weather;
- rapid run-off of water;
- acceptable riding quality;
- a waterproof seal;
- protect the base;
- an acceptable noise level;
- a pavement that stands up to economic expectations;
- maintenance-free performance for the duration of its design life; and
- adequate visibility in all weather conditions.

There are three main types of surface treatments:

1. hot-mix asphalt (premix);
2. surface seals commonly referred to as chip and spray methods; and
3. slurry methods, also a form of surface seal.

Hot-mix asphalt (HMA)

This is manufactured at a plant and despatched to site. Both the bitumen and the aggregates are heated to temperatures above 160 °C and thoroughly mixed before being loaded into a truck for transportation to site. Once the hot-mix asphalt reaches the site, it is placed into a road paver and spread onto the surface. It is important that the premix remains hot during placing. Directly after placing, it is compacted using static rollers (a pneumatic-tyred roller and a steel-wheeled roller).

There are generally three grades of hot-mix asphalt:

- **Continuously graded asphalt** is a stable, dense mix with a soft binder (80/100 penetration bitumen).
- **Gap-graded asphalt** has a certain range of particle sizes missing. Harder grades of bitumen are used (40/50 penetration bitumen).
- **Open-graded asphalt** mix has single-size aggregates. This type of mix is used to improve road surfaces' skid resistance. The grade of bitumen used is 80/100 penetration bitumen.

Surface seals

The chip and spray process refers to the surface treatment application. The chip is the aggregate and the spray refers to the bitumen sprayed onto the surface. This method is commonly known as the spray and chip method because this is what happens in reality. The bitumen is first applied either by hand spraying or using a large tank distributor, and the aggregates or chips are applied immediately afterwards followed by the compaction rollers. Several applications use this method:

- **Single seal** is used for lightly trafficked roads. The process involves applying a tack coat, spreading the aggregate and rolling with care to ensure that the aggregate does not disintegrate.
- **Double seal** is similar to the single seal except that another layer of aggregate is added. Usually the first layer that is placed contains large aggregate and the second layer comprises smaller-sized aggregate. A typical example of a double seal is using 19 mm aggregate for the first layer and following this with a 13 mm layer. The process therefore involves spraying a binder, spreading the larger aggregate and rolling, spraying a second binder, spreading the smaller aggregate and rolling.
- **Triple seal** is similar to the single and double seal except that now three layers of stone are placed. Once again, the aggregate sizes will decrease with each layer application. As an example, the aggregate sizes for a triple seal would be 19 mm, 9.5 mm and 6.7 mm. The process is as follows: spray a first coat of binder, spread the 19 mm aggregate and roll one pass of the steel wheeled roller; spray a second coat of binder, spread the 9.5 mm aggregate, and roll one pass of the steel wheel roller; then spray a third coat of binder, spread the 6.7 mm aggregate and roll.
- There are various other surface seal types and if you are interested in these, you can consult the *TRH3 – surfacing seal* document. You may obtain this document from the library, internet or your lecturer.

Slurry seals

Slurry is a mixture of fine grade aggregate (less than 4.75 mm), water, cement and bitumen emulsion. Its prime function is to fill surface voids. The slurry is prepared in a mixer on site and is evenly spread using special brooms. It is then rolled. A common application in residential areas is a product called 'Ralumac'.

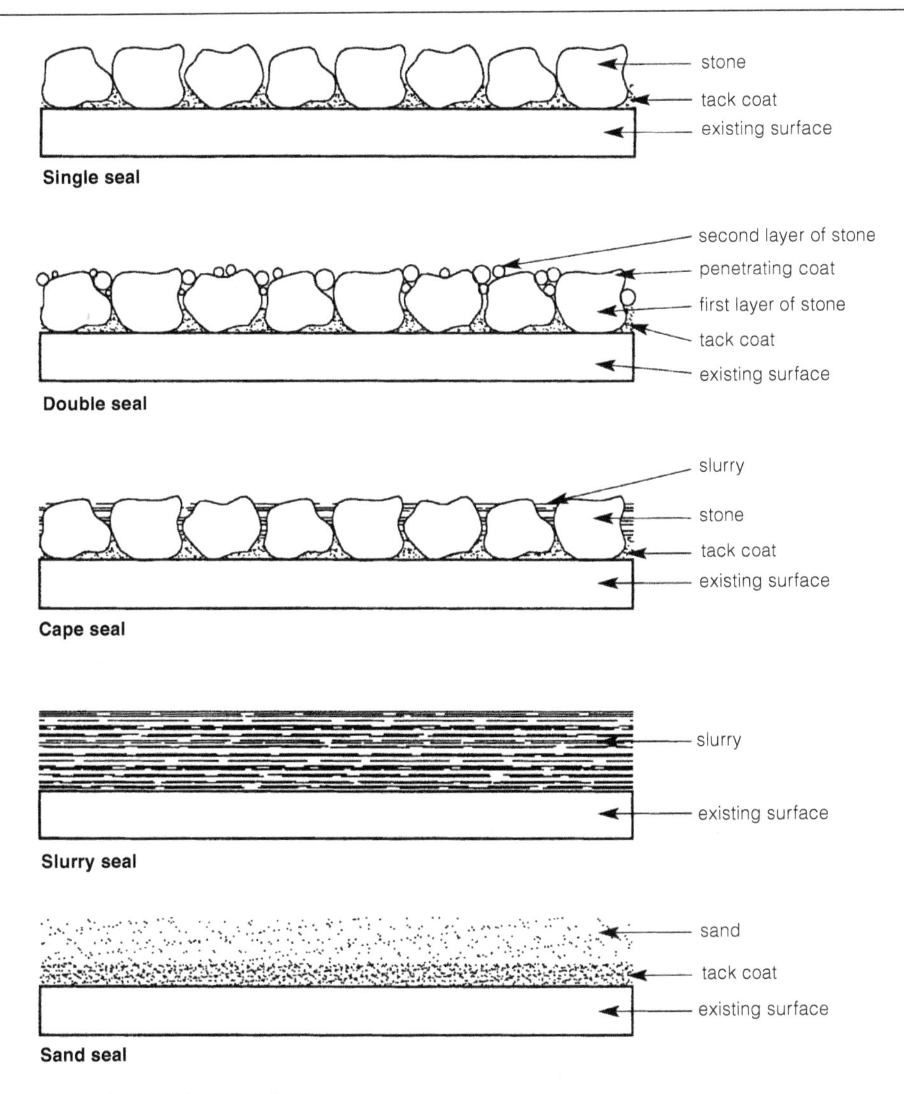

Fig 3.14 Road surface treatments

3.6 Rigid concrete pavements

This is a form of road construction using a concrete slab laid over a strong layer like a base course or a sub-base. The preparation of the subgrade is the same as that for flexible pavements. Care should be taken to prevent water from entering the subgrade. The sub-base or base course is laid over the subgrade and forms a working surface over which to cast the slab. The sub-base course layer for rigid pavements must comply with specifications and requirements similar to those for flexible pavements. The thickness of the concrete slab will depend on the nature and amount of traffic over the design life of the road, the conditions and strength of the subgrade, and whether or not the slab will be reinforced.

For a normal subgrade, using a sub-base layer thickness of 150 mm, the slab thickness will vary from 125 mm for a reinforced slab to 200 mm for an unreinforced slab carrying light traffic.

As with reinforced concrete for structures, road slabs must also conform to the 28-day strength specification, by which time the strength of the concrete should be at least 3.8 MPa. Before the concrete is laid, the base layer must be covered with a slip membrane of polythene sheet that prevents grout loss. Concrete slabs are usually laid between pressed steel road forms that are positioned and fixed to the ground using steel stakes. These side forms are designed to enable hand tamping.

Reinforcement (generally a welded steel mat) can be included in rigid pavement construction to prevent the formation of cracks and to reduce the number of expansion and contraction joints used. If bar reinforcement is used instead of the steel mat, it should consist of deformed bars (Y-bars) at spacings prescribed by the designer. The cover of concrete over the reinforcement will depend on the thickness of the concrete. For slabs less than 150 mm thick, the minimum cover should be 50 mm and for slabs over 150 mm thick, the minimum cover should be 60 mm.

Joints in rigid pavements may be either transverse (across the slab) and/ or longitudinal (along the length of the slab) and are included in the design to:

- limit the size of slab;
- limit the stresses due to subgrade restraint; and
- make provision for slab movements such as expansion, contraction and warping.

Joint spacing is governed by a number of factors, namely slab thickness, reinforcement, traffic intensity and the temperature of the air at the time the concrete is placed. Five types of joint are used in rigid road and pavement construction:

- **Expansion joints** are transverse joints at 36 to 72 m centres in reinforced slabs and at 27 to 54 m centres in unreinforced slabs.
- **Contraction joints** are transverse joints placed between expansion joints at 12 to 24 m centres in reinforced slabs and at 4.5 to 7.5 m centres in unreinforced slabs to limit the size of the slab panel. Every third joint should be an expansion joint.
- **Longitudinal joints** are similar to contraction joints and required where the slab width exceeds 4.5 m.
- **Construction joints.** A day's work should normally end at an expansion or contraction joint. If this is not possible, a construction joint can be included. These joints are similar to contraction

joints, but the two portions are tied together with reinforcement. Construction joints should not be placed within 3.0 m of another joint and should be avoided wherever possible.

- **Warping joints** are transverse joints that are sometimes necessary in unreinforced slabs to relieve the stresses caused by vertical temperature gradients within the slab if they are higher than the contractional stresses. The detail is similar to that of contraction joints, but they have a special arrangement of reinforcement.

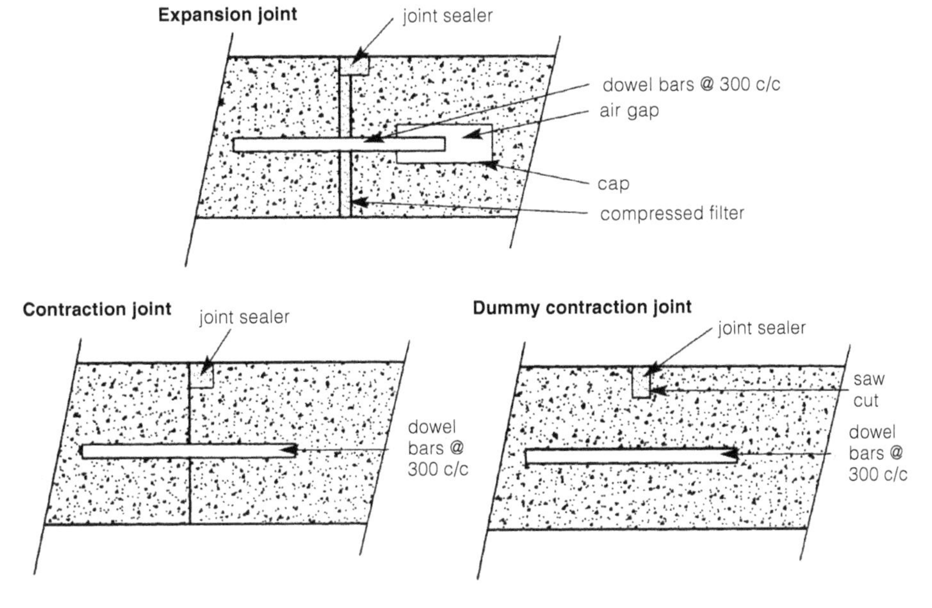

Fig 3.15 Road joints

Road joints may require fillers and/or sealers. Fillers need to be compressible, whereas sealers should protect the joint against the entry of water and grit. Suitable materials for fillers include impregnated fibreboard, chipboard, cork and rubber. The common sealing compounds used are resinous compounds, rubber-bituminous compounds and straight run bitumen. The sealed surface groove used in contraction joints to predetermine the position of a crack can be formed while casting the slab, or sawn into the hardened concrete using a water-cooled circular saw. Although slightly more expensive than formed joints, sawn joints require less labour and generally give a better finish.

The curing of newly laid rigid roads and pavings is important if the concrete strength is to be maintained and the formation of surface cracks avoided. In normal concrete construction, curing precautions must start as soon as possible after pouring, and preferably within a few minutes of completion by covering the newly laid surface with a suitable material to protect it from the rapid drying effects of the sun and wind. Light covering materials, such as waterproof paper and plastic sheeting, can be laid directly onto the surface of the concrete. Care must be taken to ensure that the ends of the covering are securely fixed to the edges of the slab form. The covering should remain in place for about seven days in warm weather and for longer in cold weather.

Activity 3.6

Prepare a table wherein you list the differences between rigid (concrete) pavements and flexible (asphalt) pavements. Speak to your lecturer, consult internet sources and speak to professionals in the road building industry to assist in compiling this table. Once complete, compare this with other tables done by your classmates.

3.7 Road drainage

Road drainage involves directing the surface water to suitable collection points and conveying the collected water to suitable outfalls. Surface water is encouraged to flow off the paved area by crossfalls, which must be designed with sufficient gradient to cope with the volume of water likely to be encountered during a heavy storm. This prevents vehicles from skidding or aquaplaning. In South Africa, road crossfalls are between 2 and 3%, which is a popular value in many road profiles. Road profiles can either be **cambered**, where the centreline of the road is the highest level and the edges taper off to either side using the specified percentage, or they can be **crossfalled**, where the road falls uniformly to one side only at the specified percentage. In most residential developments, you will find that the roads are all designed to a crossfall as it is more economical to have drainage points only on one side of the road.

Drainage, and specifically road run-offs, can also be affected by the super-elevation of the road. This would result in the cross section of the road changing to accommodate its design speed. Refer back to section 3.4.2 where super-elevation is explained. Can you think how the change in the road crossfall will affect its drainage patterns? Discuss your findings in class.

Run-off water is directed to the edges of the road, where it is conveyed by surface channels (at a minimum fall of about 1:200) in the longitudinal direction into road gullies or catchpits and then into stormwater drains.

Road gullies or catchpits are available in precast concrete form and made to rectangular sizes with either a precast concrete cover or a cast iron cover and frame. These covers are slotted to allow easy access for the water but trap road debris like litter. The spacing of catchpits is dependent upon the anticipated storm conditions and road profile, but common spacing is between 30 and 50 m. Roads that are not bounded by kerbs can be drained via subsoil drains (beneath the verge) or drain directly into a ditch or stream running alongside the road.

What do you think is done on gravel roads to control surface run-offs? Gravel surfaced roads are designed to similar geometric principles as surfaced roads, except that they do not have a surface covering. It is therefore accepted that the road profile will be either a camber or a crossfall. In order to reduce erosion and scour of the surface, more regular methods of addressing road drainage are incorporated in order to remove the water from the surface as quickly as possible. Methods such as mitres, cut-off drains, more aggressive road profiling, etc. are all methods employed to control drainage on unsurfaced roads.

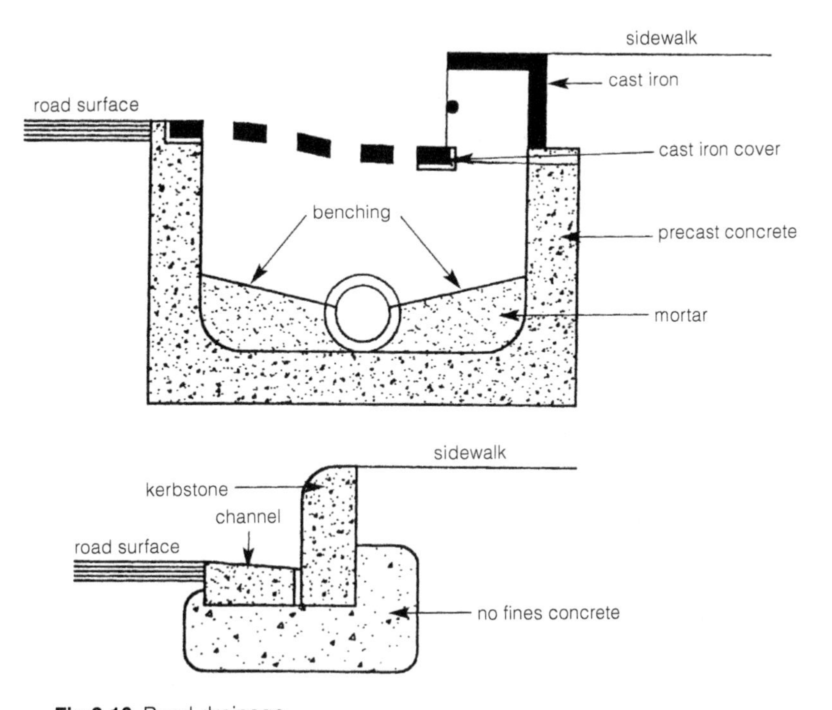

Fig 3.16 Road drainage

3.8 Accommodating services

The services that may need to be accommodated under a paved area include:

- sewerage pipes;
- stormwater pipes;
- electrical supply cables;
- water mains; and
- telephone and data cables.

In planning the layout of these services, it is essential that there is adequate coordination between the various parties concerned if a logical and economical plan and installation programme is to be formulated. Stormwater pipes and water mains are generally located under the sidewalk or to one side of the road alignment, whereas sewerage pipes can generally be found in the centre of the road. Similarly, electrical and telephone cables can also be found under the sidewalk in most instances. Services that can be grouped together are often laid in the same trench, starting with the laying of the lowest service and backfilling until the next service depth is reached, then repeating the procedure until all the services have been laid. It is essential that backfilling is properly done and is well compacted to minimise eventual settlement of the road surface.

Activity 3.7

Draw a cross-section of a major residential road and show all the major engineering services (and their location) that may be found both above and underground. Consult engineering guideline documents, other professionals and the internet.

3.9 Summary

The purpose of this chapter was to:

- Demonstrate the varying modes of transport
- Explain the factors considered when designing a road
- Review the mass haul diagram and its importance to earthwork calculations
- Explain the difference between rigid and flexible pavement designs
- Demonstrate the need to provide proper road drainage.

Self-evaluation 3

1. Complete the sentences:
   a. _______ is the movement of people and objects from an origin to a destination.
   b. _______ is the area extending from kerbline to kerbline.
   c. The subgrade fulfils the same function as a _______ for a structure.
   d. The grading of roads into a numbering system is called a _______.
   e. _______ is the removal of surplus material to reach the formation level of the road.
   f. _______ is the use of excess material to raise the level of low-lying areas.
   g. A _______ is a bituminous binder applied to a non-bituminous granular base to provide a bond between the base and the surfacing. ➤

h. Crusher run base course is a graded mix of coarse and fine aggregate including _______.
i. _______ is a general term for any paved surface.
j. _______ is a form of road construction using a concrete slab over a strong layer.
k. A _______ is placed at the end of a day's work.
l. A road is _______ when the centreline level is higher than its edge levels.

2. State whether the following are true or false:
   a. Normally a consulting engineering practice will be appointed to plan and design a road.
   b. Examples of local streets are the N1, N2 and N3.
   c. Groundwater and the water table play an important role in the design of a road.
   d. An MHD is a graphical representation of earthwork volumes.
   e. Waste refers to excess material used for embankment fills.
   f. Bulking is the decrease of material after excavation.
   g. A flexible pavement consists of a series of layers.
   h. A standard axle used in designing road pavements is taken as 2 800 kg.
   i. Tack coats are low viscosity cut-back bitumen.
   j. Reinforcement can be placed in rigid pavements to reduce cracking of the slab.
   k. Surface water should remain on the road surface as it is good to keep it wet.
   l. Road drainage can be affected by the super-elevation of the road.
3. Answer the following short questions:
   a. What elements are necessary when designing a road?
   b. Why do you think it is necessary to reach a reasonable balance between cut and fill?
   c. Why are contour maps needed for civil engineering projects?
   d. What is hot-mix asphalt?

Answers to self-evaluation 3.1

1. a. transportation
   b. roadbed
   c. foundation
   d. hierarchy
   e. cut
   f. fill
   g. pavement
   h. prime coat
   i. crusher dust
   j. rigid pavements

k. construction joint
l. cambered

2. a. true
   b false
   c. true
   d. true
   e. false
   f. false
   g. true
   h. false
   i. false
   j. true
   k. false
   l. true

3. a. see ref 3.1.3, page 76
   b. see ref 3.3, page 80
   c. see ref 3.4.1, page 87
   d. see ref 3.5.5, page 97

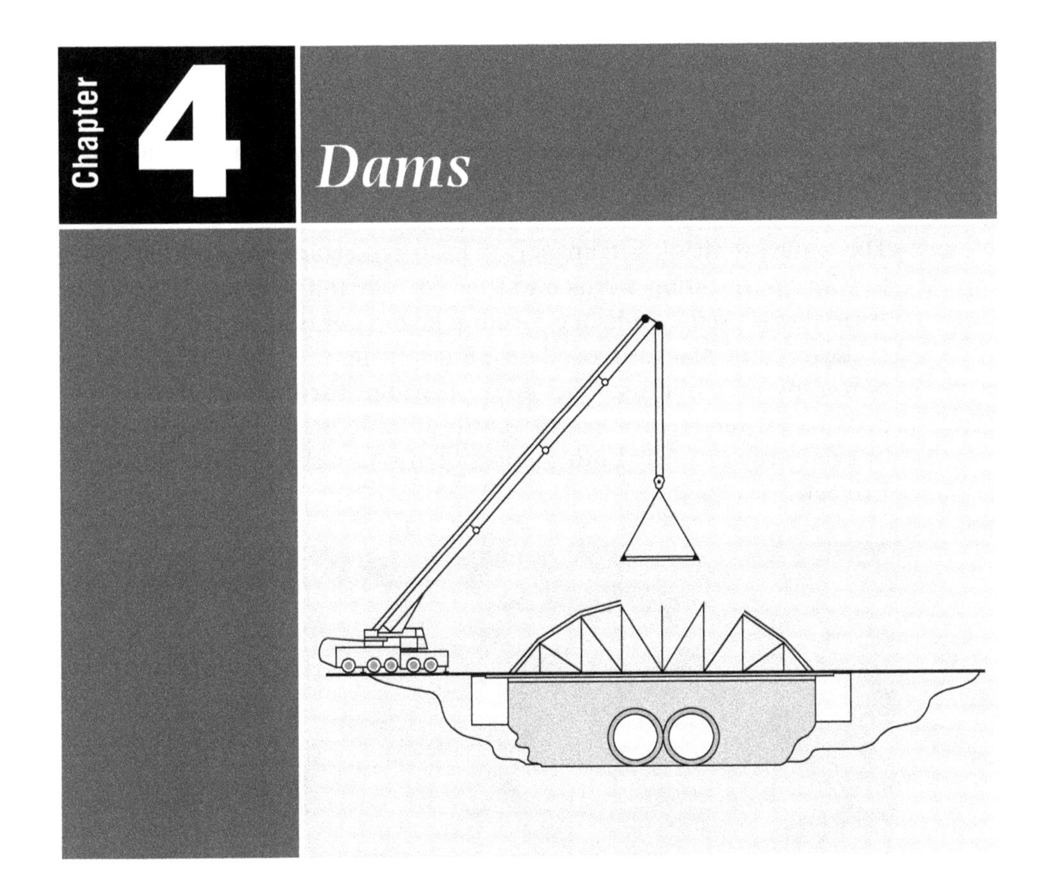

Chapter 4 Dams

Outcomes

After studying this unit, you should be able to:

- Understand what the function of a dam is
- Identify different types of dams
- Distinguish between earth dams and concrete dams
- Discuss the process involved during the planning phase of a new dam
- Explain design and construction considerations
- Discuss the importance of water conservation.

4.1 Introduction

Dams are water-retaining structures – they 'hold' water in reservoirs until it can be used at a later stage.

The water we drink is transported from reservoirs, through pipelines to water purification works and then to your tap. The water stored in reservoirs is used in industry, farming and in our homes.

Reservoirs are also used to generate power via hydro-electric power plants that use the flow of water through generators to produce power. Hydro-electric plants are uncommon in South Africa but one was recently built as part of the Lesotho Highlands Water Project.

Fig 4.1 A typical reservoir

In countries where water is exceptionally scarce – for example, in dry arid regions of central Africa and in Namibia – drain water (called grey water) is recycled for domestic use but not for drinking. The water Act makes provision for persons or companies to be fined for polluting water, particularly groundwater.

It is illegal to pour used motor oil into a gutter or a manhole. It is also illegal to pour chemicals and other harmful substances into stormwater drains as this may eventually end up in water bodies where it will pollute these.

In some countries water is so appreciated that, on rainy days, employers give their employees the day off to 'catch' the water in home-made retention facilities. What do you do when it rains? Do you give any thought to those areas where there is a scarcity of water?

If you live in cities, you will often find that the municipalities implement water conservation/restrictions, especially during the summer or low rainfall periods.

How much water do you think is lost due to leaking pipes and taps or brushing your teeth in the morning?

Can you think of innovative ways to preserve or save water? Why not start a campaign in your neighbourhood doing just that?

In all countries it is vital that some form of water retention facility is constructed to store drinking water for later use. Dams are built for this purpose, but the type and nature of the dam depends on several factors, which will be discussed later.

South Africa has always been subject to erratic and unpredictable rainfall resulting in periods of drought, often followed by severe flooding in certain parts of the country.

Also, the location of natural or mineral resources, major industrial and urban developments took place in areas where water was relatively scarce. To support these developments and a rapidly growing population, it is necessary to have a well-structured plan of water resource provision. To provide storage capacity for water (given the shortage of water during dry periods), the Department of Water Affairs and Forestry (DWAF) embarked on the construction of dams at suitable locations.

We will examine the different types of retention facilities and how they are constructed.

4.2 Definitions

- **Permeability** refers to the degree of water-tightness.
- The **stability** of a dam is its ability to resist shear forces.
- **Piping** is the constant seepage of water through a dam wall caused by water.
- **Washing of fines** is caused by wave action and is referred to as erosion of the embankment.
- **Rip-rap** is a layer of hand-placed stones (above 200 mm in diameter) used as slope protection.
- **Core** or **core wall** is the central core of impermeable material of an earthfill dam, usually clay.
- **Crest** refers to the topmost part of the embankment of the dam. It is sometimes used as a road.
- **Filter zones.** For durability, all water movements and water pressure changes within a dam have to be accommodated without any

disturbance to the fill, all of which must remain permanently and securely in place. To achieve this, drains and filters are incorporated in the dam in special zones.

- **Transition filter** is placed between the shell and core. It prevents migration of core material (usually fine soil) into pores of shell material (usually coarse grained).
- **Internal drain** prevents saturation of the upper part of the downstream shell by rain or spray falling on the dam.
- **Spillway** is like an overflow channel over a dam.
- **Toe drain** refers to the downstream toe of the shell and serves a purpose similar to that of an internal drain. It particularly prevents sloughing of the downstream face from rainwater or seepage saturation.
- **Freeboard.** All earthdams must have sufficient extra height known as freeboard to prevent overtopping. The freeboard must be of a height that wave action, wind and earthquake effects will not result in the overtopping of the dam.
- **Shell** refers to the outermost part of the dam wall that usually consists of coarse-grained material and its slopes are protected against erosion and wave action.
- **Scour** refers to the clearing of a channel by the force of water.
- **Arch dams** incorporate a curved concrete wall to span an opening and to retain water.
- **Sliding** is to move or cause to move a structure by the downward movement of a large mass of earth caused by erosion.
- **Gravity dams** use the force exerted by the mass concrete wall to retain water.

4.3 The reason dams are built

- To store water for municipal, industrial or domestic use;
- For irrigation and drainage;
- For water-oriented recreation (e.g. sailing and waterskiing);
- For recreational fishing (e.g. trout fishing)
- For flood control (retaining water to prevent large-scale damage from flooding);
- For sediment retention and control;
- For industrial waste disposal or storage (some large industries use large amounts of water in manufacturing goods);
- To produce power (through hydroelectric power plants); and
- In landfill sites, to store a substance called leachate which is very harmful to the environment, but especially to our water sources.

Activity 4.1

Can you name the dams that generate hydroelectric power? Try to find some photographs or articles to share with other students. Whatever the reason/s for building a dam, our approach must always be that a dam is one of the group of more important civil engineering works constructed by mankind for his physical, economic and environmental betterment. However, to be financially, socially and environmentally justified, the structure must be planned, designed and constructed to operate efficiently.

4.4 Dam feasibility study

Planning mainly involves the running of tests and experiments to determine a proposed dam's details and location.

4.4.1 Site selection

The site selected for the proposed dam is very important and certain factors must be considered:

- width and shape of the site – for example, in a wide-open area or a confined valley?
- potential depth in relation to storage available, yield of catchment and potential use;
- the foundations – which will determine the type of dam built;
- availability of materials for construction;
- spillway potential for non-overflow dams; and
- the prevailing wind direction.

4.4.2 Site investigation

A proper site investigation must be carried out including:

- assessment of the site topography;
- a geological evaluation (using drilling, trenching, shafts and surface investigations) of the *in situ* ground (soil or rock);
- a permeability test (using boreholes); and
- for large dams, hydraulic model tests and evaluation of existing dams of similar design.

4.4.3 Detailed investigation considerations

1. **Water supply**. This checks the availability of water, for example river, rain, groundwater or melting snow (in other parts of the world).
2. **Floodlines and stream flows.** Floods can damage dam walls and surrounding properties. Remember the floods in various parts of the country and all the damage that was caused by water. Floodlines

and stream flows, before and after floods, must be established as well as the flow rate and flooding intervals. Water is a very destructive force that, if not properly controlled, will result in disaster.

3. **Location.** This involves finding the most suitable location for the dam site. Geological factors and the availability of materials are important selection criteria.
4. **Sedimentation rates.** This is the rate at which sedimentation will settle. When this happens, sediment can settle on the reservoir floor and eventually cause 'silting up', making the reservoir shallower. The same phenomenon occurs when a river flows into a large basin like a harbour.

When referring to a **dam**, we relate to the physical structure. **Reservoir** is used to describe the body of water.

5. **Forces.** Have you noticed that water always finds its own way, especially in an uncontrolled state? It will make its own pathways and follow the path of least resistance, particularly when there is a large volume of water. I am sure you have noticed the destructive forces of a river in flood – now try to imagine controlling this with a dam wall. Can you imagine the forces acting on the dam wall? Remember that the water behind the wall will not be static but moving and wave action will further contribute towards the forces acting on the wall.

Sediment is the term for the very fine soil particles that are carried in suspension when water flows fast, but settle when the flow rate decreases.

6. **Limitations** in terms of the location due to landslides, safety of people, loss of land (where the proposed reservoir will be) and financial constraints.
7. **Environmental factors.** The disturbance of the natural scenery and ecology of the area by the reservoir must balance or positively improve the area. The reservoir must also be an asset to the fauna and flora of the area.
8. **Construction period.** It is unwise to construct or erect a dam during rainy seasons. For this reason, and because the river must be diverted in the case of on-stream dams, the construction period must be carefully planned. However, and because of the duration of construction of a large dam, it will be inevitable that construction will last for more than one or more rainy seasons. It is important to plan and make allowance for these rainy periods.

9. **The life span of the storage unit.** The life span of the proposed dam must be established during the planning stage so that the water need will be fulfilled by the storage capacity, thereby justifying the civil engineering structure economically. Imagine building a large reservoir that becomes obsolete in five years' time because it does not cater for the needs of the city or town that draws water from it.
10. **Evaporation.** Tons of water evaporate annually and not much can be done to control or stop this. Remember the water cycle? In some areas there are large natural underground aquifers containing vast volumes of ground water. The advantage of underground water is that the water is protected from the sun and does not evaporate. Have you noticed what happens to a puddle of water if it has been exposed to the sun for a day? The water disappears naturally through evaporation. Now imagine a large water body, such as a reservoir or the ocean, and consider the amount of water that evaporates on a hot summer's day.
11. **Wind, tide and possible wave action.** The maximum velocity of the wind in the area of the dam must be determined to see what the windforce on the dam wall will be, both during and after construction. Similarly, wave action can cause erosion of the dam wall, which must be guarded against.
12. **Earthquake resistance.** Where possible, dams must not be located in areas prone to earthquakes or in the vicinity of faults in the earth's crust.

4.5 Dam types

In practise, there are many types of dams – home-made ones, farmers' irrigation dams and large retention facilities that supply our daily water. Generally, dams can be classified as:

4.5.1 Embankment dams

- **Earth dams** of silt, clay, clay and sand, sand and gravel.
- **Rockfill dams** comprising rocks only.

4.5.2 Concrete dams

- **Simple arch dams** are suitable where the height/length ratio is less than 1:4, otherwise little economic advantage over gravity dams is gained. They require a sound rock foundation in both the floor and the sides of the valley.
- **Buttress dams** are only suitable for wide stream beds where shuttering can be repetitive and, therefore, economical.

Do you still remember what timbering is and how it is used? If not, look again at Chapter 1.

4.5.3 Primary design considerations

Any retaining structure must be designed to prevent **sliding and overturning.** To prevent this happening, specialists design the structure taking into account the forces such as water pressure, wind, waves, soil pressure, etc.

4.5.4 Secondary design considerations

Dams should also be designed to prevent or withstand:

- **Overtopping** through careful siting, with spillways, sluice gates and scour valves. Overtopping happens when a dam becomes too full and water then flows over the top of the dam wall. You will find that all dams are built with a fairly substantial difference in level between the natural water level in the dam and the top of the dam.
- **Scour** (erosion, usually on the downstream side).
- **Wave action** on the upstream side.
- **Leakage** in various forms including rodent action and piping (tubular voids caused by seepage pressure).
- **Earth movements.**

4.6 Embankment (earth) dam design

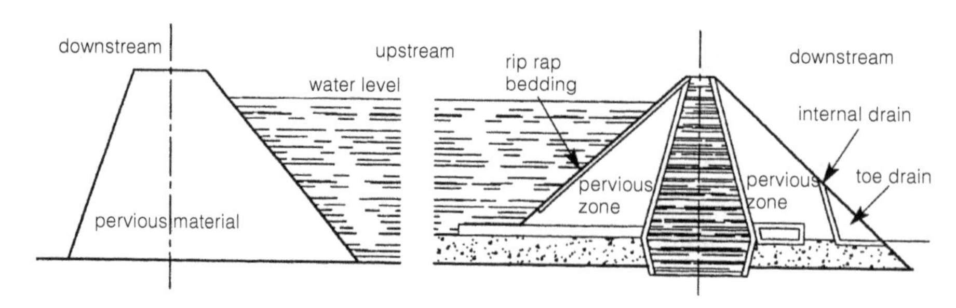

Fig 4.2 Sections through embankment dams

An example of an earth dam is Nooitgedacht Dam on the Komati River. In this dam, which is constructed entirely from soil, the height of the dam wall is 44 m above the lowest foundation. The volume of earth used was 1 330 000 m^3.

How many truckloads of 6 m^3 capacity would have been required to move this amount of earth?

Recent advances in soil mechanics have resulted in earth dams being considered as safe as any other kind of dam. Earth dams were the first type of dams to be constructed and now cost about half the price of a concrete dam in comparable circumstances. Generally, material for earth dams is easily available and no specialised labour is required. Some of the largest dams are made of earth and, in India, some are even centuries old. The main challenge in designing earth dams is to select a composition of materials and side slopes so that, with a given foundation condition, no failure of the dam will occur. Since the function of the dam is to hold back water and all earth is permeable to some extent, we may anticipate that the problem of slope stability is intimately associated with the problems of water movement through and under the dam.

4.6.1 Design components

- **Core** or **core wall** is the central core of impervious material in an earthfill dam. The material used is often clay.
- **Cut-off core** or **cut-off wall** is the continuation of the core taken as deep as is necessary into the ground so that contact is made with impermeable rock. The cut-off core protects the foundation from under-seepage. It can be the same material as the core or it can be different (normally clay or concrete). A deep cut-off core consists of a narrow trench excavated across the valley below the dam. This trench, about the width of a bulldozer, is cut through all permeable strata lying near the surface of the ground and penetrates into the sound, watertight rock at the base. It is backfilled with clay or concrete.
- **Crest**. The top part of the embankment of a dam. It is normally 3.0 m wide and is often used as a road.
- **Earth embankment**. The ridge of earth built up to contain the water in the reservoir.
- **Filter zones**. For durability, all the water movements and water pressure changes within the dam have to be accommodated without disturbing the remainder of the embankment, which must remain permanently and securely in place. To achieve this, drains and filters are incorporated in the dam in special zones.
- **Transition filter**. This is placed between the shell (the outer edge) and the core. It prevents the migration of core material (usually fine-grained soil) into the pores of shell material (usually coarse-grained soil).
- **Internal drain**. This prevents saturation of the downstream shell.

- **Freeboard.** All earth dams must have sufficient extra height, known as freeboard, to prevent overtopping by the water. The freeboard must be of sufficient height that wave action and earthquake effects will not result in overtopping of the dam.
- **Overfall/overflow/overspill.** The part of a dam over which the water pours.
- **Rip-rap.** Large aggregates or concrete blocks needed to cover the upstream face of the dam to prevent erosion or wash by waves.
- **Spillway.** Overflow channel.
- **Toe drain.** Downstream toe (the bottom of the downstream slope) or shell serving a similar purpose as an internal drain. It particularly prevents sloughing of the downstream face from rainwater or seepage saturation.

4.6.2 Design considerations

1. **Permeability.** This is the rate at which soil will allow water to pass (seep) through it. Material that is impermeable – for example, rock – has a very low seepage rate, whereas permeable material – for example, sand – has a higher seepage rate. No earth or rock dam is completely impermeable, but the flow of the water through the dam wall must be controlled by using various construction materials.
2. **Stability.** The stability of a dam is its ability to resist shear forces. Forces on the dam may cause it to collapse. These include water pressure, wind and wave action, foundation and soil pressure, and forces within the dam itself.

Activity 4.2

Let's try the following experiment. Use a large steel or glass bath and partition it so that you can create a barrier filled with various types of material. One third of the bath must be 'constructed' with the material listed below, while the other third must be filled with water to emulate a dam scenario. The material to be used must be the following:

- 4.6 mm stone
- Clean building sand
- Laterite or gravel with a high clay content.

This experiment will be done three times, each time using a different material from those listed above, which must be compacted into one third of the space allocated. The purpose is to observe the rate of seepage though the makeshift wall. Which material has the lowest permeability? Which material has the highest porosity?

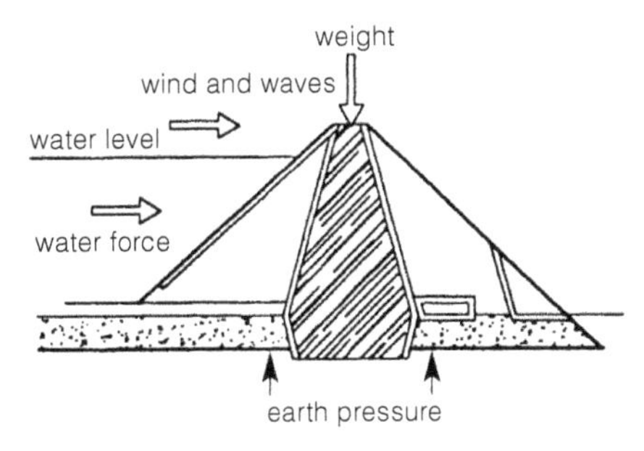

Fig 4.3 Forces on a dam wall

3. **Compression and shrinkage.** A dam will sag under its own weight, mainly because voids occur between the particles of the construction material during construction. Silty material cracks when dry, therefore the moisture content of the construction material after erection must be kept constant.
4. **Washing of fines.** Washing is internal erosion of fines (washing out of fines from the interior of the soil caused by seepage). Washing of fines is also caused by wave action and is referred to as embankment erosion. A layer of large aggregates with a minimum diameter of 200 mm used as rip-rap (very large aggregates that are closely hand-packed on a slope) is used as slope protection.
5. **Availability of construction materials.** If construction materials for a proposed dam can be gathered on site or in the vicinity of the proposed dam, this will be very advantageous for the economy of the structure.

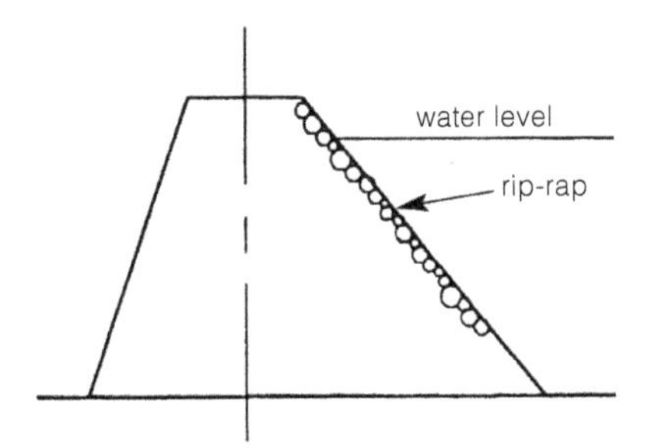

Fig 4.4 Rip-rap on an earthfill dam

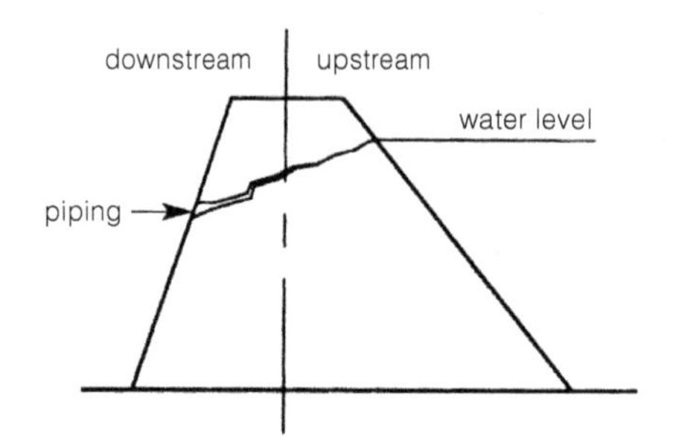

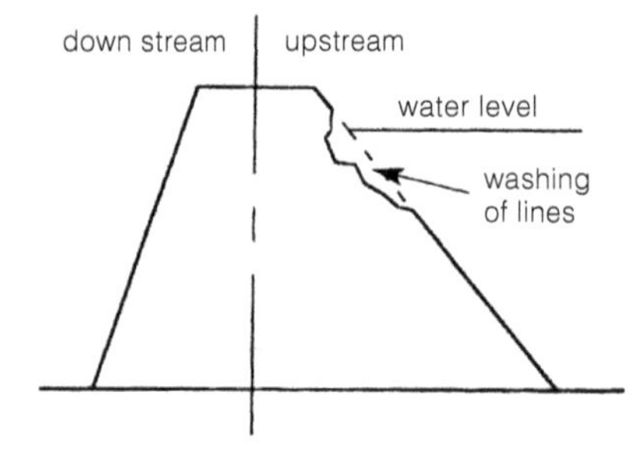

Fig 4.5 Seepage through dam walls

4.6.3 Hydraulic aspects of embankment dams

When a dam of homogeneous material is situated on an impervious and stable foundation and permanently exposed to a head of water, certain seepage through the dam will occur.

The seepage pattern and the place where the seepage line emerges on the downstream slope is the same whatever material the dam is made of, provided the material is homogeneous and all dimensions are similar in each case. The rate of seepage depends entirely upon the type of soil used and the head of water. For example, through a dam composed of fine sand, seepage is considerably more than a similar dam composed of clay. Why do you think this is so? Remember, when dealing with soils, we stated that clay has properties that tend to retain water and decrease its permeability. Well, this is still true in dams where the clay reduces the amount of water that seeps through the embankment. The emergence of seepage lines on the lower part of the dam has a tendency to make the slope less stable than it would be under dry conditions. So, either a relatively mild slope that will be stable while the seepage emerges needs to be designed, or the seepage must somehow be diverted from the downstream slope.

The first option would be rather costly, but the second is more feasible and widely applied in dam design. One way of preventing the seepage lines from emerging on the downstream slope is to design the dam with a low permeability core and high permeability shell or toe fitter.

The small amount of water that seeps through will flow in layers off the downstream fill and emerge at the toe of the dam where a gravel and rock filter is located.

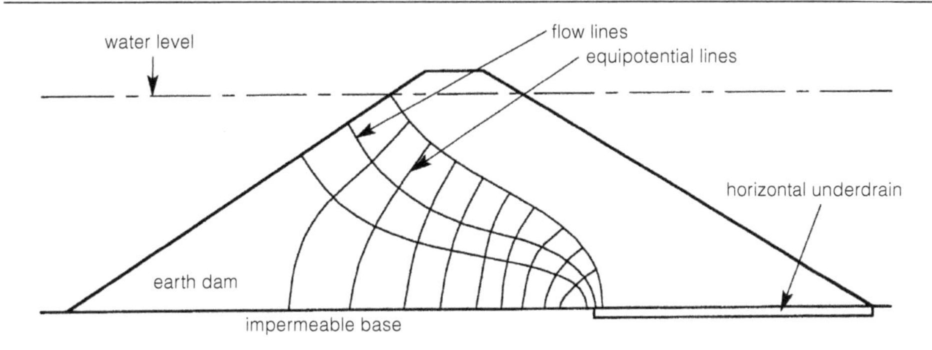

Fig 4.6 Flow through homogeneous earth dams

4.6.4 General principles

There are a few general principles that, if followed, ensure a safe and durable structure. The first and most important consideration in earth dam construction is that the dam should be made as near impermeable to the action of water as possible. Seepage water is destructive in many ways, of which the following are most important:

- The saturation of the dam imparts an effect of buoyancy that decreases the effect of gravity upon which, to a large extent, the stability of the dam depends.
- Water acts as a lubricant and causes the soil particles to slide over one another. This is a common cause of failure.
- Seepage may occur at any point along a dam. Once started, the volume of seepage water increases rapidly and results in erosion.

4.6.5 Freeboard

The difference in elevation between the top of the dam wall and the maximum level to which the water in the dam may rise is called the freeboard. The freeboard may be as little as a few metres, for a gravity dam holding back a small reservoir, to as much as 8 m in large reservoirs made of earth.

Adequate freeboard is necessary since the overtopping of the dam by water flowing over it would be disastrous for the stability of the dam. An earth dam with a saturated downstream slope may fail because of progressive sliding. A gravity or rockfill dam resting on an earth foundation or keyed into earth embankments has the potential to fail at the toe or at the flanks of the dam if overtopping occurs.

4.6.6 Rip-rap

The protection of the dam from wave action can be accomplished in many ways. If rock is available, rip-rap can be placed on the upstream side of the dam, at that part exposed to wave action.

Rip-rap should be placed in a bed of sand or gravel so that the voids in the rock may be filled, to prevent the action of the waves reaching the earthen part of the dam. Cement and bitumen are sometimes used to fill the openings between the rocks. A much better option, if possible, is to locate the dam so that the prevailing wind blows away from it, and the waves do not wash the dam slopes unduly.

4.6.7 Homogeneous dams

A homogeneous dam is made from the same kind of material throughout, offering consistency in all its parts. If there is any difference in the grade of material used, the better or more impervious material should be placed on the upstream side of the dam. If impervious clay is unavailable, the material should be mixed and graded.

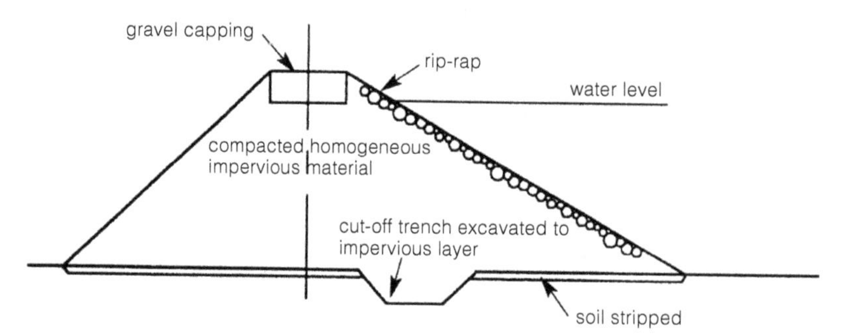

Fig 4.7 Homogeneous dam

4.6.8 Zoned dams

This is an earth-filled dam that has a core wall in the centre. The core wall is made from a mixture of clay and sand (called puddle earth) which creates an impervious barrier to seeping water.

This type of dam is very susceptible to earthquakes during its construction. Until the dam wall has dried out, the pore pressures are very high. The dam wall often dries out gradually. It is also accepted that settlement of the embankment will take a long time.

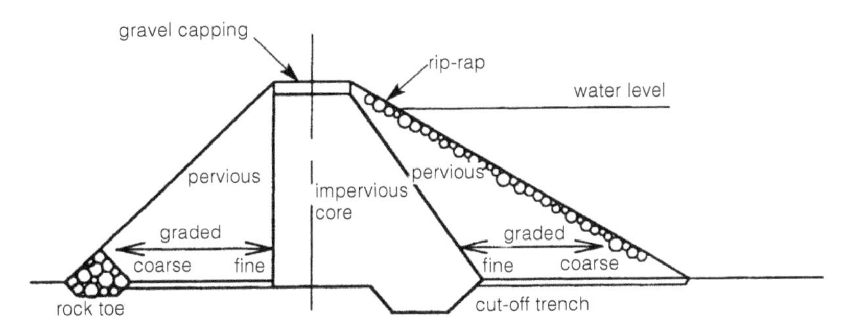

Fig 4.8 Zoned dam

4.6.9 Impervious water-faced dams

This type of dam is used when there is sufficient impervious material available to build one-third to one-half of the dam structure. Material is placed on the upstream side and considerable care is taken in its placing. The balance of the dam or downstream side is built of coarser material that will supply the necessary stability that comes from the mass. It has an additional advantage in that it will offer good drainage should there be any seepage. This type of dam is normally constructed as a rolled fill dam.

4.6.10 Advantages of embankment dams

1. The materials are generally **readily available**. Modern advances in soil mechanics have made it possible to use previously unsuitable material.
2. Materials are **easily workable**, both manually and with large machines.
3. Movements in the foundations could cause cracks in concrete dams. Embankment dams are more **adaptable** to such movements.
4. Embankment dams are generally **cheaper** than comparable structures made of concrete.

4.6.11 Disadvantages of embankment dams

1. If suitable material is unavailable near the site, transport can raise costs due to the large quantities required.
2. Considerably more maintenance is required.
3. A special overflow is required.

4.7 Concrete dam design

4.7.1 Gravity dams

The Gariep Dam, built on the Orange River, is an example of a gravity dam. It has a minimum height above foundation level of 90.5 m and had a volume of 1.73 million cubic metres of concrete. The Steenbras Dam in the Western Cape is also an example of a gravity-type dam wall structure. In this case, the height of the wall is 36 m above foundation level and it had 51 000 m^3 of concrete poured. To support a wall of this height and the mass of concrete, it is absolutely essential that it has a solid foundation. The foundation must take into account:

- the horizontal thrust due to the water pressure;
- the weight of the dam wall acting downwards; and
- any upthrust from the ground due to water pressure.

Considering all these factors should ensure that the dam wall remains stable and is safeguarded against overturning. If the riverbed has a poor rock foundation, an apron is constructed downstream to prevent erosion that could undermine the dam foundations through backscour.

?

Why do you think concrete dam walls are curved? Which is the best direction to curve the dam wall (i.e. convex or concave) and why?

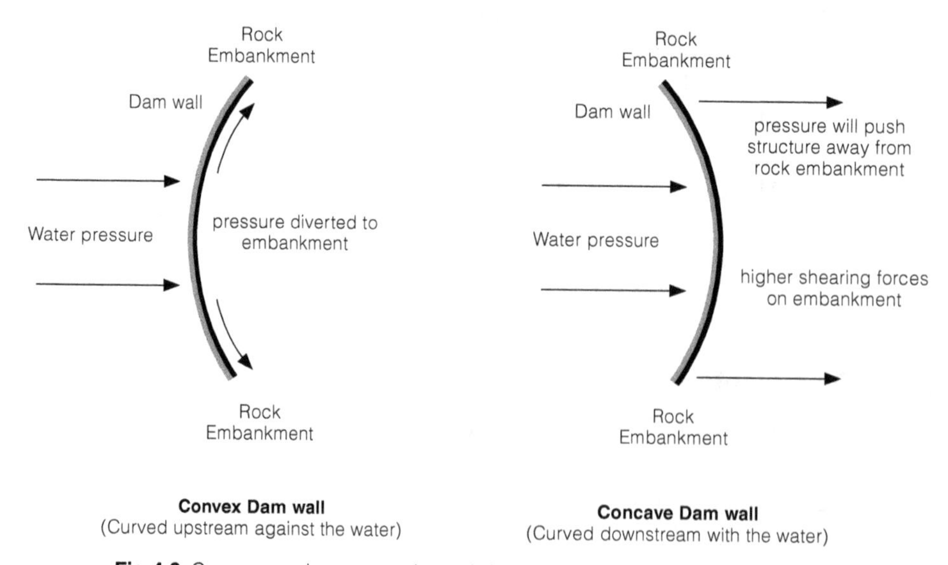

Fig 4.9 Convex- and concave-shaped dam walls

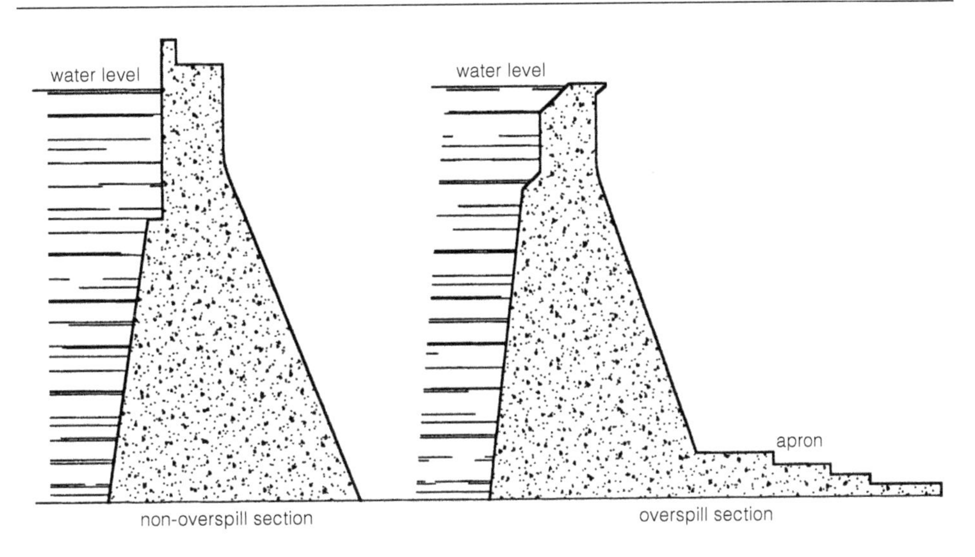

Fig 4.10 Concrete dam walls

4.7.2 Buttress dams

An example of this type of dam is the Inga I on the Congo River and the Rhenosterkop Dam in South Africa. These dams have a sloping upstream face and the weight of the water above the face adds to the vertical forces. Thus less concrete can be used than for gravity dams. The cost of shuttering is considerably increased and the saving in cost is minimal unless concrete and labour are relatively cheap. To counterbalance the massive increase in the cost of shuttering, the dam may have to be built on a concrete foundation if the base rock is poor.

Fig 4.11 The Rhenosterkop dam is an example of a buttress dam

4.7.3 Simple arch dams

Examples of this type of dam are Bangala Dam in Zimbabwe, the Gariep Dam and Roode Elsberg Dam. The main design criterion here is that the water pressure is transmitted to the banks, although the vertical cantilever effect must also be taken into account. These dams require very sound foundations and failure of parts of the dam wall means failure of the dam. The design is fairly complex and the construction is usually somewhat more expensive per cubic metre than that of gravity dams. For fairly wide valleys, gravity sections are used that are curved in plan to provide additional safety. If the dam is designed to overflow, provision must be made to reduce the effect of vibrations by using ski jumps to dissipate the energy.

Fig 4.12 The Gariep Dam is an example of a simple arch dam

4.7.4 Multiple arch dams

In multiple arch dams – for example, Wagendrift Dam – the horizontal thrust of the water is transmitted to the buttresses.

Fig 4.13 The Wagendrift Dam

4.8 Dam construction

The design and construction of large dams is a specialised field of civil engineering. It is also a team effort involving civil engineers, mechanical and electrical engineers, environmentalists, geologists, hydrologists and other supporting personnel.

The design and construction of each large dam is a challenge since no two sites are the same. Various types of dams exist, designed to solve problems peculiar to their sites.

4.8.1 Rock-fill dams

Rock-fill dams consist mainly of compacted rock-fill. Rock-fill is usually obtained from a quarry located close to the dam. Ideally, the quarry site is chosen so that, on completion of the dam, it can serve as a spillway. The rock is blasted in the quarry, hauled to the dam by 35 ton or bigger dumper trucks, spread in layers of between 800 and 1 500 mm by a bulldozer and compacted with heavy vibratory rollers.

The rock-fill zones of a dam are not watertight – in fact, they are free draining. To retain water within the dam basin, it is necessary to provide an impermeable zone or a membrane. This can be achieved by two methods. The rock-fill embankment can be lined on the upstream slope by either a concrete or an asphalt lining, or an impervious zone of clay can be included in the centre of the embankment. When a clay core is used, specially graded filters of sand and crushed stone have to be used to ensure that the clay core is kept in position and is not washed away.

The Mokolo Dam (Hans Strydom Dam) is an example of a rockfill dam with a central clay core. This dam is located on the Mokolo River about 45 km upstream of Lephalale in Limpopo. The dam supplies water to a coal mine and a large power station near Lephalale, and irrigation water to downstream users.

In this particular case, the foundation, which consists of hard quartzitic Waterberg sandstone, was favourable for the construction of either a concrete or a rock-fill dam. Due to the good quality of the foundation rock and the relative scarcity of suitable earth-fill material close to the dam site, an earth-fill embankment dam was not considered.

Cost analyses proved that it would be slightly cheaper to construct a rock-fill embankment than a concrete dam. It was decided to use a central clay core because, although suitable clay material was relatively scarce, it would still be cheaper than an upstream concrete membrane. The Department of Water Affairs was also able to gain valuable experience in the construction of rock-fill embankments, that could be applied to other large structures being planned at the time of construction.

The construction of the dam started with the blasting of a large diversion tunnel through the foundation rock on the left riverbank. After the tunnel was completed and lined with concrete, a cofferdam was constructed across the river and the river was diverted through the tunnel. The main dam was then constructed. The intake tower at the mouth of the diversion tunnel was constructed at the same time as the main embankment, and a temporary gap was left between the tower and the tunnel inlet. On completion of the main dam, the temporary diversion opening was closed and two outlet pipes were installed in the tunnel. The diversion tunnel thus became part of the normal outlet works of the dam and serves as access to the valves in the intake tower.

The spillway has a capacity of approximately 12 000 m^3/s and is located on one side of the dam wall. The quarry for all the rock-fill used in the construction of the dam wall served as the spillway channel.

Can you calculate how many litres of water equate to 12 000 m^3?

4.8.2 Earth-fill embankment dams

These dams consist of compacted earth-fill. Earth-fill embankment dams are usually constructed where foundation conditions are inadequate to support a concrete dam.

The engineering properties of earth, as it appears in nature, are dependent on its clay content. Clay is defined as those particles smaller than 0.002 mm. The higher the clay content of a soil, the more watertight it will be, but, unfortunately, very flat embankment slopes are then required to achieve stable dam wall slopes. More sandy earth will be more pervious to water but, with its material stability, a comparatively steeper slope can be used.

Large earth-fill embankment dams are usually constructed with a centrally located clay core zone. This core zone is supported on either side by pervious zones of more sandy materials, which allow more stable slopes. A smaller total volume of embankment materials is thus required.

No earth-fill material is completely watertight and water pressure on a dam will cause seepage through the structure. Seepage water on the downstream slope can cause erosion. Wet earth-fill is also less stable than the dry material.

To keep the downstream zone of a dam dry, an internal sand or sand/gravel filter system is provided in the form of a horizontal blanket drain on the foundation and a vertical chimney drain downstream of the central clay core. This drainage system collects the seepage water percolating through the clay core of the dam and its foundations, and provides a controlled discharge of the water through the downstream toe drain, which ensures that the downstream supporting zone is kept dry. The downstream embankment slope can thus safely be made steeper than the upstream slope.

Goedertrouw Dam

The Goedertrouw Dam on the Mhlatuze River near Eshowe in KwaZulu-Natal is a large earth-fill dam. The dam provides water for irrigation, and for urban and industrial development at Richards Bay and Empangeni. During the September 1987 floods, the reservoir, which was only 58% full when the heavy rain started, absorbed a large volume of floodwater.

The foundations on the left flank (looking in a downstream direction) of the dam site were of good enough quality to accommodate a large river diversion tunnel. On the right flank, the rock had decomposed to clay to such a thickness that it would have been totally uneconomical to excavate through the clay down to the fresh rock that could support a concrete dam.

An earth dam was judged to be the most economical for the site, because a large volume of suitable earth-fill materials was present in the dam basin. The spillway was located in a natural neck about 200 m from the dam on the right flank. The spillage flows back to the river via a natural channel downstream of the dam.

Once the river diversion tunnel was completed, a 25 m high earth-fill cofferdam was built across the river and the river was diverted through the tunnel. Construction of the main embankment then started and became a race against a flood overtopping and breaching the cofferdam. At one stage, 420 000 m^3 of earth was placed and compacted in a single month.

An 83 m high reinforced concrete intake tower at the entrance of the diversion tunnel was built at the same time as the main wall. A temporary gap was left between the tower and the tunnel entrance for the river to flow through. When the main dam had progressed to a few metres above the level of the spillway crest, the temporary gap was closed off and the outlet pipes were installed in the tunnel and became part of the outlet works of the dam.

4.8.3 Concrete dams

The essential differences in the design of a concrete dam compared to an earth dam are:

- The design can be more certainly and precisely determined, despite the advance in the knowledge of soil mechanics that furthered the construction of earth dams.

- The construction materials are more stable.
- Outlet pipes can be built into the structure and the inlet tower can be part of the dam wall.
- Floodwaters may pass over the dam during and after the construction period without endangering the structure.

Another advantage of a concrete dam is that the application of a vertical force to a gravity dam design by means of a set of steel cables, enables the engineer to reduce the cross-sectional area, with a subsequent reduction in concrete.

Concrete dams can be divided into different types of dams, namely masonry, gravity, arch and buttress dams or a combination, such as an arch-gravity dam.

Masonry dams

This type of dam is built of stone bedded in mortar reaching either over its full width or on the outside faces only. No use is made of shuttering and the final results can be very aesthetically pleasing. The use of masonry facing and the practice of embedding large stones have become obsolete and these dams are seldom built these days.

Gravity dams

The stability of this kind of dam depends on its weight and the density of the concrete, which should be as high as possible. As concrete is weak in tension, the tensile stresses at any part of the dam must be very low under any loading condition. As concrete gravity dams have to remain economically competitive, new, cheaper construction methods have been developed which have led to the use of roller-compacted concrete or rollcrete.

Zaaihoek Dam

Zaaihoek Dam is one of the first rollcrete gravity dams to be constructed in South Africa. The main purpose of the dam is to supply water to the Majuba Power Station near Volksrust. Water will also be released for industrial agricultural and domestic users downstream of the dam. Water pumped from the dam to the Vaal River catchment will increase the yield of the Grootdraai Dam which supplies most power stations on the Eastern Highveld.

The dam is situated on the Slang River in the Tugela catchment about 12 km southwest of Wakkerstroom. Upstream of the dam, the foundation rock changes from sandstone to dolerite. The softer, more erodible sandstone provides a good wide reservoir basin, while the hard dolerite provides a strong foundation for the dam structure. The dolerite forms a relatively narrow gorge, making the site ideal for a concrete structure. For economical and topographical reasons, and for speed of construction, the dam was of a rollcrete gravity construction.

Rollcrete, or roller-compacted concrete, not only describes the type of concrete, but also the method used to place it. Rollcrete is basically a stiff concrete designed to be transported by equipment such as dump trucks and conveyer belts. It is spread by bulldozer in layers about 300 mm thick across the length and width of the dam, and compacted by vibratory rollers. Rollcrete is placed without contraction joints that are utilised in conventional mass concrete. This method makes the construction of the dam wall much easier and faster, and leads to time and cost savings.

The downstream face of the dam had to be adapted to suit this construction method and a stepped face was used for Zaaihoek Dam. This also helps to dissipate the energy of the water discharged over the spillway, and a smaller apron slab at the toe of the dam was needed to protect it from the scouring action of the water. The spillway of the dam is situated in the river section and is ungated, while the scour and the river outlets that release water for downstream consumers are on the right-bank flank. A separate set of pipes transports water from the dam to a pumping station, from where water is pumped to users in the Vaal River catchment.

Buttress dams

A massive buttress dam resembles a gravity dam in shape, but has cylindrical portions removed from the wall at regular intervals. This leaves buttresses that transmit the water pressure to the foundation. The upstream face is usually flattened to utilise the direct weight of the water in the dam, to help stabilise the structure. As the cylindrical cuts extend all the way down to the foundation, the uplift on the dam is substantially reduced. This type of dam requires a rock foundation of good quality as the stresses on the foundation are higher than for an ordinary gravity dam because they are concentrated under the buttresses. The initial design for this type of dam requires relatively simple calculations, later followed by more refined analysis.

Shuttering for this type of dam adds to the cost. It is possible that recent advances in the construction of rollcrete dams will make rollcrete gravity dams with few joints and minimal shuttering more competitive in price than massive buttress dams.

Rhenosterkop Dam

Rhenosterkop Dam, on the Elands River near Groblersdal in Gauteng, is a massive buttress dam. It supplies additional water to part of Gauteng and makes provision for water supply to a coal mine.

The dam consists of a massive buttressed spillway and non-overspill crest portions where the height is substantial. The lower ends of the non-overspill crest on both sides, and the outlet block, are of the simple gravity type. Provision has been made in the design to raise the wall by four metres. Piers were provided on the spillway at 15 m intervals, where gates can be installed at a later stage.

Various alternative types of structures were considered. A rockfill dam could have been built at a comparable price, but the shortage of clay in the area for the clay core made the rock-fill alternative less attractive. A conventional gravity-type dam would have been more expensive. A rollcrete dam would have been a suitable alternative, but this method was unknown in South Africa at the time of construction.

The possibility of building a dam at Rhenosterkop without a spillway, relying on the outlet works and the capacity of the basin to absorb floodwater, was also considered.

The Rhenosterkop Dam was the first massive buttress dam to be built in South Africa. A similar dam was subsequently built in Namibia

Arch dams

An arch dam relies on its wall arching across the valley it spans to transfer the thrust of water to the foundations of its flanks. An arch is an efficient structural element and hence relatively thin sections can be used, allowing substantial savings in concrete compared to, for example, a gravity dam. The structure may be a single or a double curvature arch, that is arched in the horizontal direction only or in both the horizontal and vertical directions. The limiting factor in the design of arch dams is the width-to-height ratio, which must be relatively small, otherwise problems can be encountered and an arch becomes impracticable. The foundations of an arch dam also have to be good, as forces exerted by the structure are large and concentrated. The geology is very often the deciding factor in choosing an arch dam design.

In many cases a combination structure relying for its stability partially on its own weight and partially on arching is used and is known as a gravity-arch or arch-gravity dam, depending on which is dominant. An example of an arch-gravity structure is the Gariep dam.

Gariep Dam

The Gariep Dam is the key structure of the Orange River Project. The dam serves several purposes:

- To supply water to the drought-stricken but fertile Great Fish and Sundays river valleys via the Orange Fish Tunnel.
- To generate hydro-electric power.
- To control floods and regulate flows to allow better utilisation of the Orange River water.

The effect of the Gariep Dam and the Vanderkloof Dam (PK le Roux Dam) on the 1988 floods in the Orange River was substantial.

Apart from delaying the flood, allowing for the complete evacuation of low-lying areas days before the flood reached them, the flood peaks were drastically reduced by the flood absorption of the dams. Far more catastrophic damage would have occurred had the dams not been there.

The dam is situated in the gorge at the entrance to the Ruigte Valley, near Norvalspont, and is an arch-gravity structure. Because the gorge at the site where the dam was built was too wide to allow a complete arch, only the central section is arched. Gravity sections on each flank form artificial abutments for the arch. The length-to-height ratio of the dam is large for an arch dam and a special heel structure is incorporated to counteract the shear effect associated with such a large ratio.

A combination of several factors led to the choice of this type of structure. A rock-fill dam could have been considered if suitable material had been available. A further factor was that, for a river the size of the Orange, a very substantial concrete spillway would have been required, as the overtopping of such a wall by a flood could lead to its failure with disastrous consequences. In the case of a concrete structure, overtopping may cause some damage, but not failure.

The choice of the combination structure was based primarily on economic reasons. The double curvature section required less concrete than a gravity section would have done, with a consequent reduction in cost.

As erodable soils are present in the Orange River catchment, siltation is a problem and scour outlets are included in each flank. In addition, river outlets are present on both banks. Pipes feed water to the 320 MW hydroelectric power station. The dam has a central spillway without gates and a gated chute spillway on each bank. The latter are used mainly for flood control.

Large cofferdams had to be used during the dam's construction to keep the water out of working areas. The dam wall was divided into a number of blocks or monoliths along its length to allow for shrinkage, thereby controlling cracking on the concrete. The contraction joints between the monoliths were later grouted by pumping in a cement/water mix under pressure through pipes installed in the concrete. Chilled water was also circulated through pipes in the concrete to reduce temperature build-up and related cracking which develops due to the heat generated during the setting process of the concrete. It also brought the concrete down to the minimum ambient temperature so that the contraction joints could be grouted.

Multiple-arch dams

This type of structure consists of a series of arches or cupolas supported on either side by buttresses. It is used where the width-to-height ratio of the valley is too large for a single arch, and where the foundations are sufficiently competent. In a multiple-arch dam, the thrust of the water is transferred via the buttresses to the foundation rock as compared to an arch dam, where the thrust is transferred directly to the foundation. As the designer can decide upon the length of each arch, within certain limits, it is possible to optimise to a degree to obtain the most economical design. Normally double curvature arches are used, which are arched in both the horizontal and vertical directions. These behave structurally as domes, which are very efficient structural elements and can consequently be very thin. It should be noted that sophisticated structural analyses are normally required for the design of these structures.

The major advantage of multiple-arch dams is that the volume of concrete required is substantially less than for most other alternatives, such as gravity or buttress dams. The disadvantage is that sophisticated construction techniques must be used. This is relatively labour intensive, requiring skilled personnel. Nowadays this factor affects costs to the extent that this type of structure is seldom used, as construction equipment has developed so that large volumes of material can be handled rapidly and cheaply. Consequently, simpler structures, though requiring larger volumes of material, tend to be more economical. Another problem with this type of structure is that water flowing over the spillway drops freely onto the foundations downstream of the wall and special precautions have to be taken to avoid erosion.

Wagendrift Dam

The Wagendrift Dam is a multiple-arch dam where double curvature cupolas were used.

The Wagendrift Dam, on the Bushmans River near Escourt, KwaZula-Natal, regulates the river flow and provides a reliable source of irrigation water to the area downstream, up to its confluence with the Tugela River.

The dam consists of four thin-walled cupolas or domes supported by five buttresses. The spillway, which is uncontrolled, is situated in the two central arches. Controlled water releases to the river are made via a sleeve valve situated in the central buttress.

Various alternative structures were considered for this dam. An earth or rock-fill dam may have been considered if suitable materials had been available. The excellent quality of the foundation rock also made a concrete dam the obvious choice. As the valley is too wide for a single arch, and taking concrete costs into account, the multiple-arch structure was more economical than a gravity or buttress dam.

This type of structure was innovative at the time of its design and construction, particularly when one considers that none of the sophisticated computerised methods of structural analysis was available then, and the analysis had to be based on hand calculations and structural model tests. The dam can rightly claim to be the first non-reinforced multiple-arch dam constructed in the world.

4.9 Water conservation

Water is a very scarce commodity worldwide and is becoming even more so due to the inconsiderate and irresponsible manner in which water is used. Water is a non-renewable resource. Once it has gone, it cannot be replaced. What will happen in 40 years' time when there is only a limited amount of water available?

There have been talks of towing large icebergs from the poles to supplement the water shortage. This may prove to be very uneconomical and dangerous.

Research is currently underway to try and convert seawater to a quality acceptable for consumption. But again, cost is a major factor.

Did you know that there are some areas in South Africa that recycle the water flushed from toilets? After going through a purification process, the water is pumped into the ground and then drawn up through boreholes to replenish drinking water.

Another water conservation practice that has become acceptable throughout the world is to recycle domestic water. When you wash the dishes, or bath or wash your clothes, the water takes on a greyish colour. This is commonly called "grey water" and can be used for various other purposes. It is not fit for drinking, but can be used to water the garden, flush toilets, etc. This is called recycling. Various systems are being researched to find the most suitable and economical means of recycling water.

Can you guess how much water a toilet cistern takes? On average, a normal toilet cistern uses 14 litres of water for every flush! Now make a rough guess how much water you use when you take a bath. Depending on how full you fill it, approximately 70 litres of water is used for a bath.

Did you know that the average consumption of water by a city dweller is 140 litres per person per day, which includes drinking and bathing? For some people this can be more, for example bathing or showering more than once a day, filling the bath to the brim, etc. Can you think of people you know who do this? Do you think people living in lower income areas use less water than those who live in higher income areas? Now take a city the size of Cape Town that has an average of 2.5 million residents and multiply this number by the average consumption per person. It's an astronomical figure, isn't it? This is not even taking into account water used for irrigation (commercial farming) and industries. Now imagine the quantity of water needed on a daily basis. What size of reservoir is needed to supply water to Cape Town? Cape Town is fortunate that it is supplied by five reservoirs located outside the greater metropolitan area, but those are large volumes of water that must come from somewhere.

Activity 4.3

In groups, discuss the following:

- areas in your daily life where you waste water;
- what you can do to help conserve water; and
- ways of recycling water.

Gather as much information as you can about any aspect of water conservation and make a poster as well as a report (of no more than 10 pages) to present to the rest of the class.

4.10 Summary

The purpose of this chapter was to:

- Describe what a dam is and how it functions
- Explain and illustrate the various types of dams in South Africa
- Explain the processes involved during various cycles of dam planning, design and construction
- Demonstrate the importance of conserving water.

Self-evaluation 4

1. Complete the sentences:
   a. ____________ refers to water that is fit or suitable for drinking.
   b. _______ is the term used to describe the very fine soil particles carried in suspension when water flows fast.
   c. _____________ are suitable for wide stream beds where shuttering can be repetitive and economical.
   d. _______ is the internal erosion of fines.
   e. The difference between the top of the dam wall and the maximum level to which the water in the dam may rise is called ______.
   f. An _____ is made from the same kind of material throughout.
   g. _______ consists mainly of compacted rock-fill.
   h. _______ is defined as those particles smaller than 0.002 mm.
   i. No earth-fill material is completely watertight and water pressure on a dam will cause _______.
   j. Buttress dams require a _______ foundation of good quality as the stresses on the foundation are high.
   k. '____________' is the term used to describe water from households.

➤

2. State whether the following are true or false:
   a. Dams are built to go sailing on.
   b. The sun influences the amount of evaporation that takes place.
   c. Crest refers to the topmost part of a dam wall.
   d. Rock has a low permeability.
   e. Water acts as a lubricant and causes the soil particles to slide over one another.
   f. Embankment dams are cheaper than comparable structures made of concrete.
   g. Gravity dams depend on their weight and the density of the concrete for stability.
   h. Rockfill embankment dams are more expensive to construct than concrete dams.
   i. Concrete dams can be designed more accurately and with more certainty than embankment dams.
   j. An arch dam transfers the thrust of the water to the foundation of its flanks.
   k. The volume of concrete required for multiple arch dams is more than that of traditional shapes.
3. Answer the following short questions:
   a. Name the authority that controls water usage in South Africa.
   b. Why is it necessary to undertake detailed investigations before designing a dam?
   c. What is the difference between a dam and a reservoir?
   d. What important considerations need to be taken into account when designing dams?
   e. List the differences in design approaches for concrete dams and earth dams.
   f. Why is it important to conserve water?

Answers to self-evaluation 4

1. a. potable water
   b. sediment
   c. buttress dams
   d. piping
   e. freeboard
   f. homogeneous dam
   g. rockfill dams
   h. clay
   i. seepage
   j. rock
   k. grey water
2. a. true
   b. true

c. true
d. true
e. true
f. true
g. false
h. true
i. true
j. false
k. true

3. a. see ref 4.1, page 108
   b. see ref 4.3, page 110
   c. see ref 4.4.3, page 111
   d. see ref 4.4.2, page 111
   e. see ref 4.7.3, page 124
   f. see ref 4.9, page 133

Chapter 5

Bridges

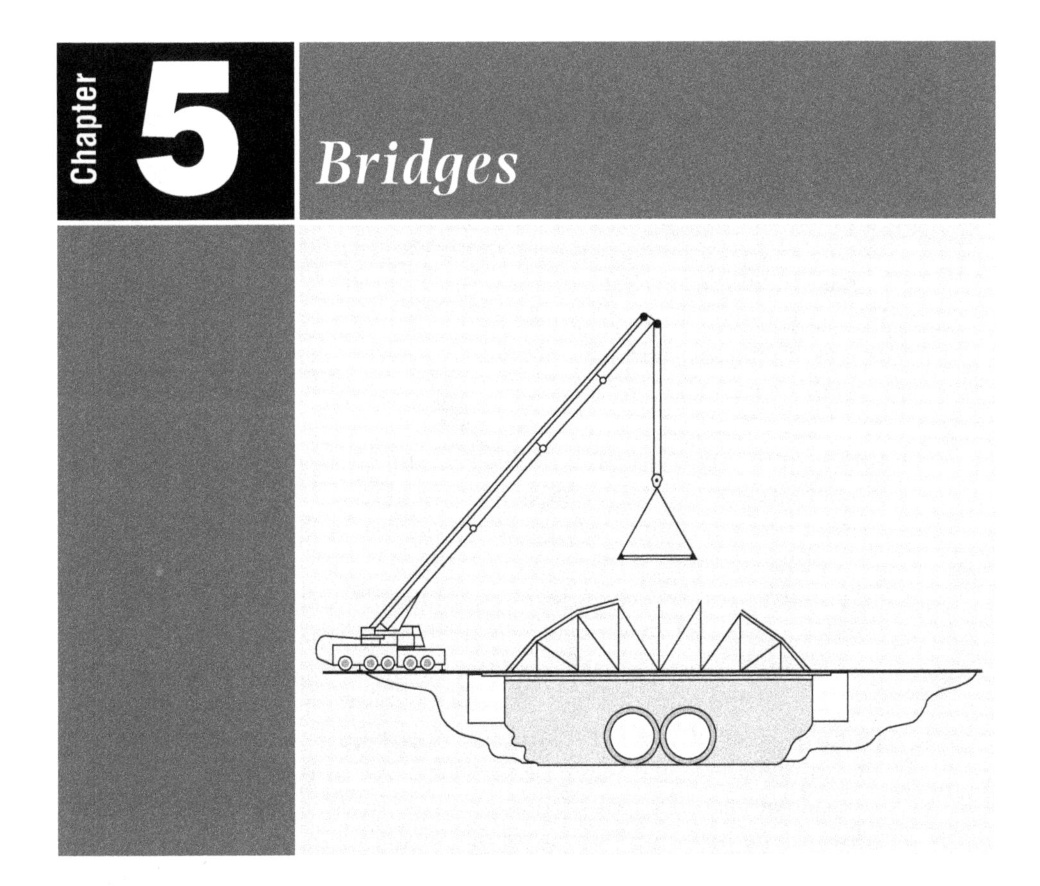

Outcomes

After studying this unit, you should be able to:

- Explain bridges and their design principles
- Distinguish between the different types of concrete and steel bridges
- Identify the component parts of bridges
- Discuss the role each part plays in the whole structure.

5.1 Introduction

A bridge is a structure that enables a service, such as a road, a railway line or a pipeline, to pass over another service or obstacle without disrupting either. In its simplest form, a bridge consists of a beam or beams supporting, for example, a road, over a span.

Fig 5.1 Old stone arch bridges

Think about a bridge structure in your environment and its functions. Now imagine what people would have done if the bridge was not in place.

Ever since the first tree fell across a stream, bridges have been part of the natural landscape. Almost every day we cross bridges, be it by foot (at a railway station), in cars and in trains (e.g. the Bloukrans Bridge on the Garden Route), over rivers (e.g. the George Washington Bridge, New York), canals (e.g. in Amsterdam, the Netherlands) and roads. The most popular examples of bridges are associated with our road networks. Think of some of the world famous bridges and their link to transport, for example the Golden Gate Bridge, Brooklyn Bridge, etc.

In days past, the polished stone axe was developed and used for cutting and trimming logs to make simple beam bridges. Before modern materials like steel and concrete were discovered, bridges were constructed from stones.

The first bridges were made of straight beams supported at each end. Irrespective of the material used, builders found that there was a limit to the distance these could span. This was because each material has a certain tensile strength and fails if the span is too large.

Bridge engineers then discovered that, if they used an arch, they could span much larger distances because they could make use of the compression strength of the materials. Do you remember from your Construction Materials course that stone and concrete have much higher compressive strength than tensile strength? Also go to the following websites to extract more information relating to bridges: www.pbs.org/wbgh/buildingbig/bridge/ as well as http://science.howstuffworks.com/engineering/civil/bridge.htm

Today, with the use of materials like steel and reinforced concrete, bridges can be built with considerable spans using either straight beams or arches. Bridges represent the combined skills and technology of architecture, design and engineering.

Did you know that the first bridge built in South Africa was completed in 1892, (on the road between Cape Town and Mossel Bay). This bridge was 220 m long and 60 m high.

5.2 Types of bridges

There are various types of bridges, some of which use very sophisticated methods of construction.

5.2.1 Arch bridge

An arch bridge is a curved structure (in the form of an arch) constructed in the vertical plane and used to span an opening. You will often find arch bridges where there is a need to cross a steep valley or where it would be too costly to support the bridge at intervals along the span.

An arch exerts a thrust sideways on its foundation. It is like bending a plastic ruler into the shape of an arch. Feel the pressure exerted on your fingers as the ruler tries to straighten itself. This pressure is similar to the sideways thrust experienced on bridges.

Different methods can be used to construct arches. For example, an arch can be built out from both sides of the valley to meet in the centre, supported by cables over towers at each end (abutment). When the two sections of the arch meet, the cables are removed and the road can then be constructed, supported by columns resting on the arch. An example of this type of bridge is the Bloukrans Bridge.

Another method involves building each half of the arch on each side of the valley and then lowering the two sections with cables and cranes until they meet at the centre where they are joined. An example of this type of bridge is the Storms River Bridge. Imagine how accurately and precisely you have to construct these halves in order to make them join perfectly once placed together.

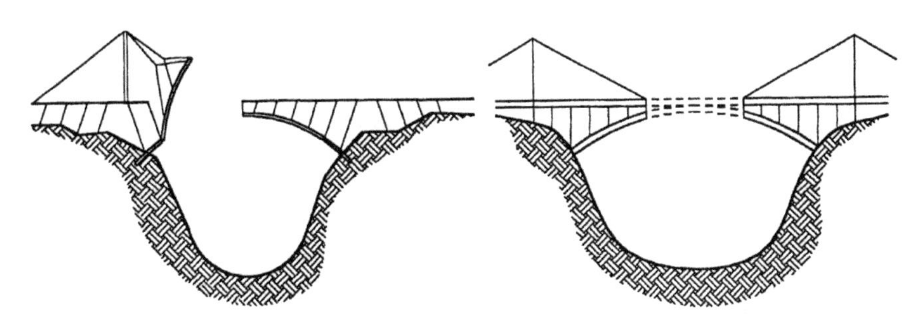

Fig 5.2 Arch bridges

5.2.2 Suspension bridge

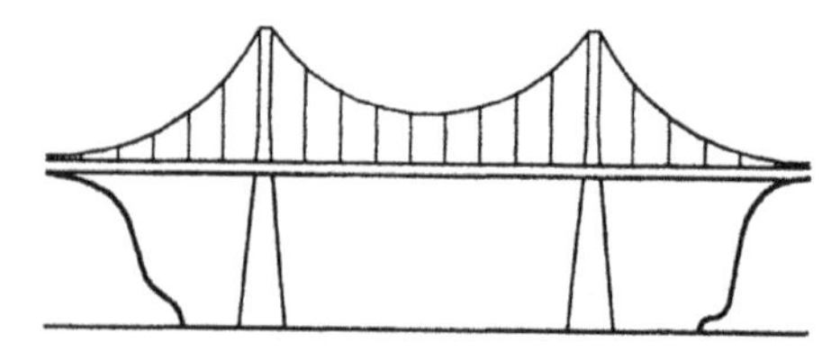

Fig 5.3 Suspension bridge

One of the most famous suspension bridges is the Golden Gate Bridge in California. Suspension bridges are used where very long spans are needed. They consist of tall towers on each side of the span, with two main steel cables spanning between them. The road is then suspended from the main steel cables using hangers.

The Golden Gate Bridge has two main cables, 920 mm in diameter, which pass over the tops of the main towers and are fixed at each end. The length of one cable is 2 332 m.

The main cables of the Forth Road Bridge in Scotland are 700 mm in diameter and span a distance of approximately 1 500 m. Can you imagine a cable this thick and long? Imagine the equipment and manpower needed to handle a cable of this size.

5.2.3 Cantilever bridge

These bridges are constructed from both sides of the valley to meet at the centre, both sections cantilevering over the support. Do you still remember what cantilevers are? If you have forgotten, refer to your Drawings and Theory of Structure subjects.

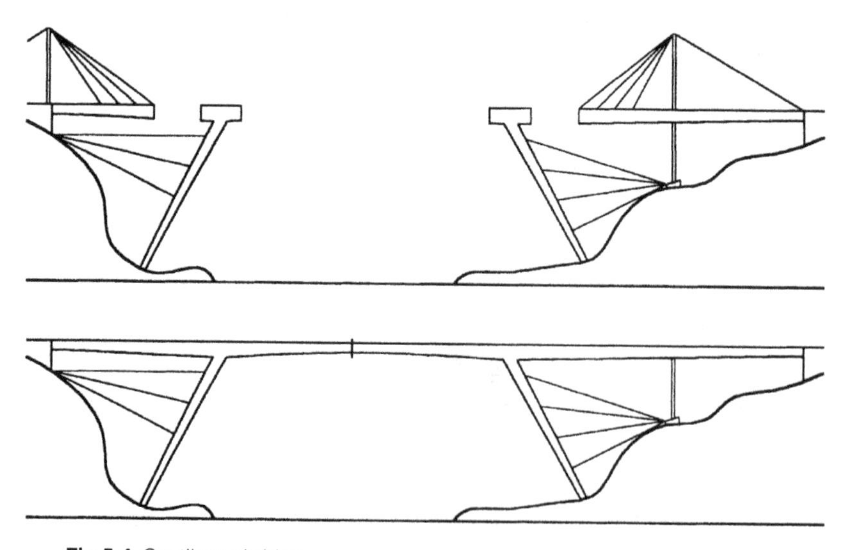

Fig 5.4 Cantilever bridge

5.2.4 Incremental launching bridge

In this type of construction, the entire bridge deck is built in sections on one side of the span. A shutter is fixed, as each section has a steel nose that spans forward to rest on the piers (pillars) constructed to support the deck.

The next section is then cast in the shutter and joined to the first using reinforcing bars. Post-tensioning cables that are threaded through ducts cast in each section provide additional stability.

As each section is completed and pushed out of the shutter, the bridge deck is extended to span over the piers. When the deck has reached the opposite abutment, the steel nose is removed and the rest of the work is completed.

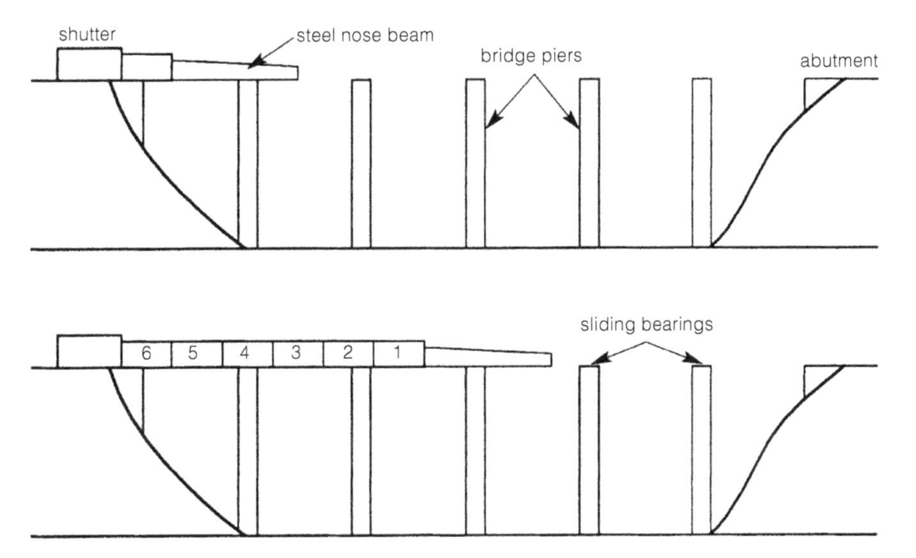

Fig 5.5 Incremental launching bridge

This type of construction must ensure a high degree of accuracy with little margin for error.

5.2.5 Cable-stayed bridge

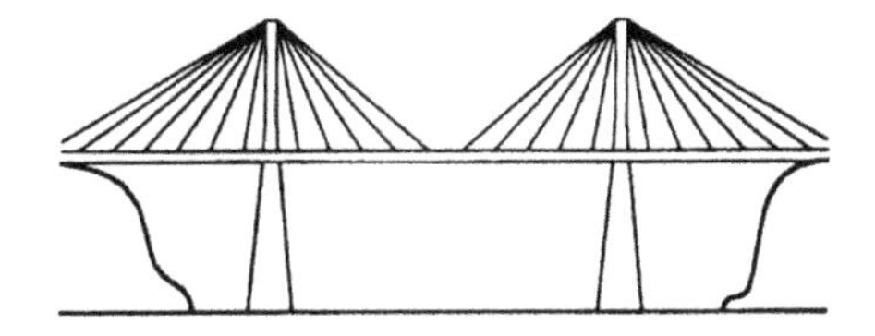

Fig 5.6 Cable-stayed bridge

This type of bridge uses cables that are directly attached to the tops of the towers or piers. The road is then suspended from these cables. Two or more towers, each with its own set of cables, supports a section of roadway.

5.2.6 Draw bridge

A draw bridge is used over a waterway where there is insufficient clearance between the water and the underside (soffit) of the bridge.

The deck is hinged at the supports near the banks and made in two sections that can be raised and lowered – for example, when a ship wants to pass through. London's Tower Bridge is an example of this type of bridge.

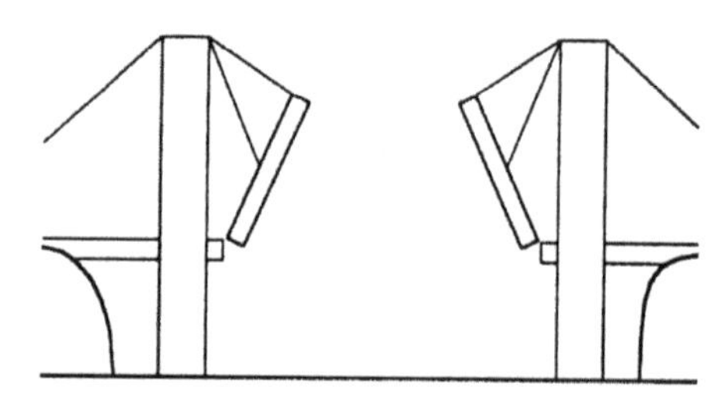

Fig 5.7 Draw bridge

5.3 Bridge design components

5.3.1 Bridge abutments

For the purpose of this chapter, an abutment to a bridge may be simply defined as the support to the end span at an embankment or cutting of the bridge deck. There are two types of abutments: **wall abutments** and **open abutments**, the latter are also known as **bank seats**. Wall abutments serve two purposes – supporting the deck and retaining the embankment or cutting – whereas bank seats provide support only to the bridge deck.

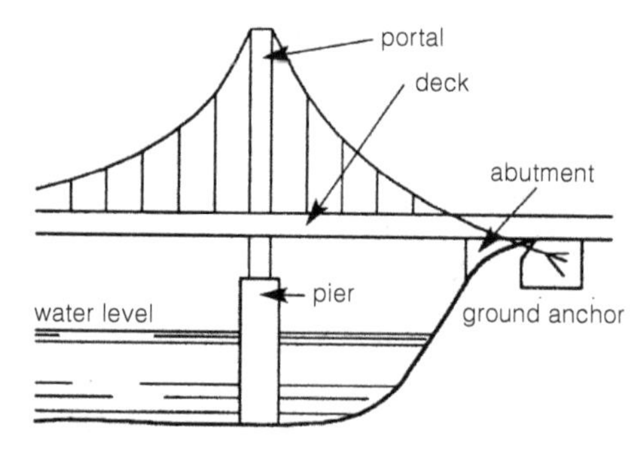

Fig 5.8 Bridge elements

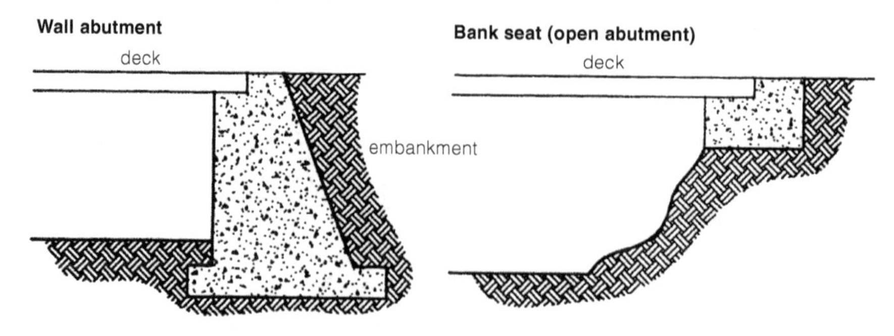

Fig 5.9 Bridge abutments

Wing walls

Wing walls are used to either complete the visual aspects of a bridge abutment or to form an integral part of the bridge – for example – when the walls are parallel to the overhead road. The walls are usually constructed in reinforced concrete although crib type (used in special cases) may be positioned parallel, angled or at right angles to the line of the abutment.

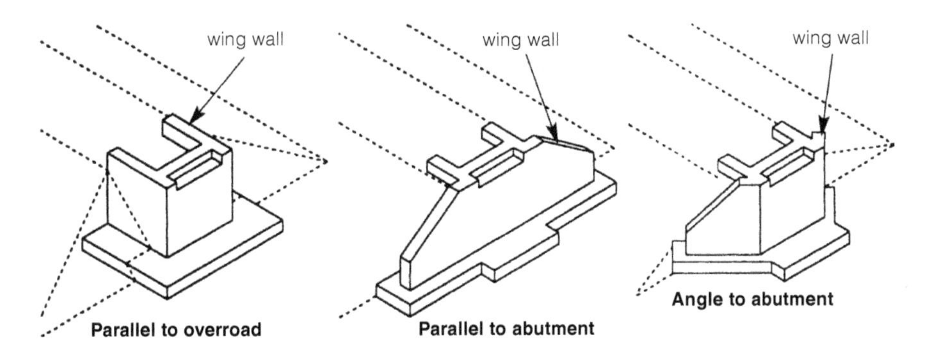

Fig 5.10 Wing walls

Crib-type wing walls refer to a framework of heavy timbers laid at right angles to one another.

5.3.2 Piers

Piers, like abutments, support the bridge deck but occur within the length of the deck rather than at the ends. The simplest form of pier is a vertical member of uniform cross-section although, for aesthetic reasons, piers are often of non-uniform section and have a textured surface.

If the deck of the bridge has a capacity to span transversely, column-type piers at suitable centres may be used. Where no such capacity exists – for example, in the case of box beam bridges – either solid reinforcement concrete wall piers the full width of the

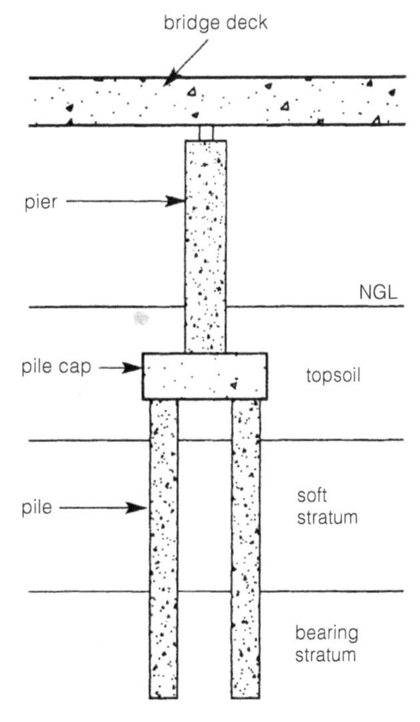

Fig 5.11 Typical bridge pier

deck or cross-head portal frames in reinforced concrete or steel would normally be adopted.

When single column-type piers are founded on piles, although the load capacity may suggest a single large-diameter pile, it is more practical to use a group of smaller diameter piles. The reason for this is that problems could arise if the pile is inaccurately set. To allow for the inspection of the reinforcement and any cleaning out that has to be done prior to pouring the concrete, piers should have a sufficiently large cross-section to allow a man to climb inside if possible. This will also facilitate the placing of the concrete. Raked piers and 'V' piers are often used – the latter in particular as the deck spans are increased in comparison with vertical piers, and in both cases for aesthetic reasons.

Do you remember reading about pier foundations in Chapter 1?

5.3.3 Superstructure

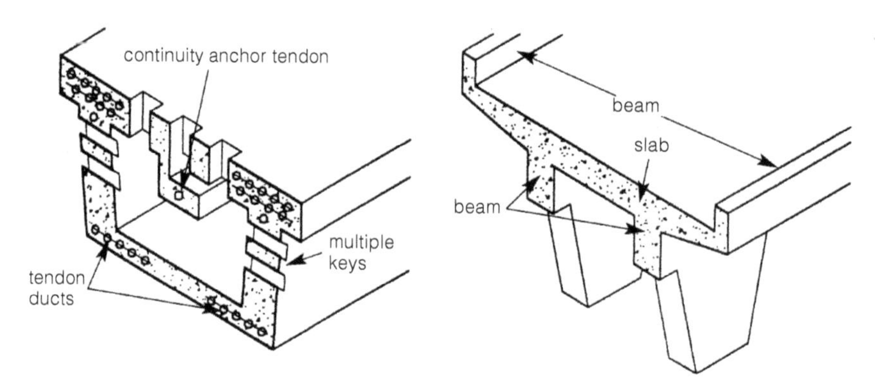

Fig 5.12 Bridge deck

Fig 5.13 Beam and slab

This is the deck of the bridge, which rests on the abutments. Many rectangular-type bridges have a considerable span and require deck 'stiffening'. This can be achieved by 'beam and slab' construction.

5.3.4 Bearings

Up to three types of movement may be allowed for on a bridge bearing: horizontal movement, longitudinal or transverse movement, and rotational movement. For spans of less than 16 m, provision is normally only made for horizontal movement. For greater spans, angular rotation of the deck at its joints due to traffic loading must also be accommodated.

Why do think it necessary to provide movement of the structure on bearings?

The simplest form of bearing is a **rubber strip** or **pad**. This type of bearing is used to resist vertical loading and, to a limited extent, rotational movement. **Elastomeric bearings** comprising steel plates laminated with vulcanised rubber offer similar performance to the simple rubber bearings, but also permit horizontal movement by shear displacement.

There are several structural bearing types used in bridge construction today:

Pot-type bearing

This is a generic reference to the concept of a structural bearing in which an elastomeric (non-reinforced) disc is encapsulated in a 'pot' and 'piston' configuration of metallic components. This causes the elastomeric disc to act like a fluid when a load normal to the surface is applied. The disc allows rotational movement between the members to occur, resulting in a very small rotational movement in the structure.

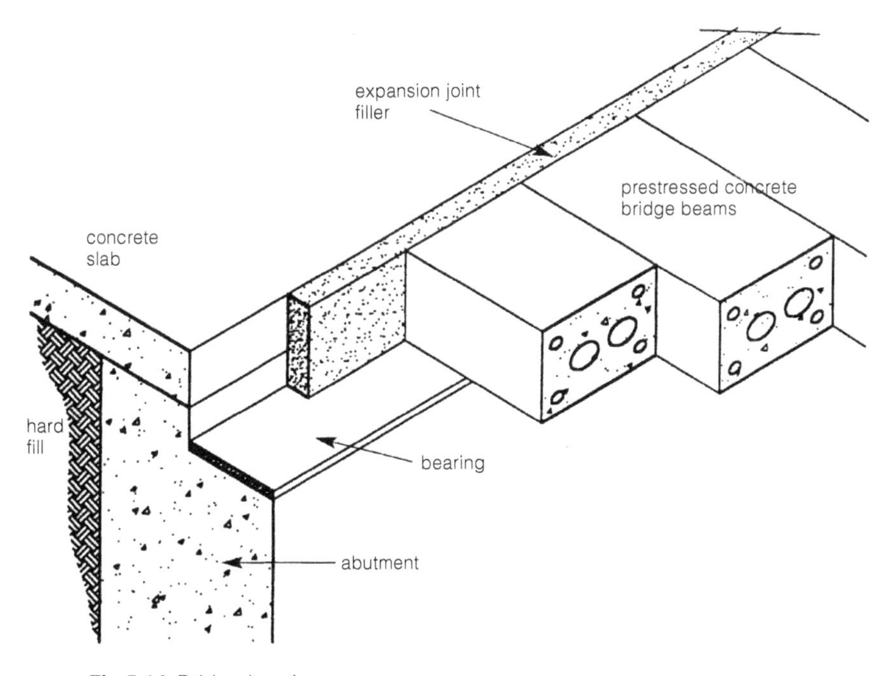

Fig 5.14 Bridge bearings

Elastomeric bridge bearing

This type of bridge bearing refers to a non-reinforced bearing pad or strip used in civil engineering applications to support either concrete or steel superstructures. It makes use of a simple yet effective separation strip or pad that can carry compressive loads while, at the same time, provide transverse movement.

Why is it necessary to cater for movement in a bridge structure?

Structural slide bearing

As the name suggests, this type of bearing either floats or guides/slides movements within the structure.

Slip-joint bearing

This is designed to support *in situ* concrete where it provides for the centralisation of the load transfer together with a low coefficient of friction to reduce stress in the concrete during shortening. These bearings are typically used in multi-level car parks to provide for rotation induced by the settlement of imposed loads.

Rocker-type bearings

These are 'older' types of bearings and can be thought of as springs on which structures are mounted. They are manufactured from cast steel.

5.3.5 Expansion joints

Read about expansion joints under 'Concrete' in *'Construction Materials'*.

Expansion joints in bridge decks are used principally to accommodate horizontal expansion and contraction of the deck. This, in turn, requires provision to be made for the accompanying movement of the deck parapets, crash barriers, kerbs, etc. The number of joints needed will depend on the predicted maximum movement and the capacity of the joints to take up that movement.

However, most expansion joints interrupt the road surface and, therefore, reduce the quality of the ride. And because of the continuous impact of traffic, they tend to need fairly regular inspections and maintenance.

5.4 Criteria for designing a bridge

Certain properties must be taken into account when designing a bridge, including:

- the normal load to be carried;
- wind loads;
- the span;
- the foundation (the type of subsoil and its condition):
- the height of the structure (will it clear flooding?); and
- the siting of abutments, piers, etc.

Material that is strong in tension is important in bridge building. Do you remember what types of material are strong in tension?

The only part of an arch bridge structure that is not in tension is the arch – the rounded or semi-rounded portion that spans an opening.

Steel is an invaluable material for bridge building. The Golden Gate Bridge in California is a famous example of a cable suspension bridge. Steel is mostly used for suspended bridges, because the bridge hangs from steel cables. The cables, suspended between two towers, are all in tension. This type of bridge is used for long spans.

5.5 Bridge materials

The different bridge forms are constructed mainly of three distinct materials:

- steel;
- reinforced concrete; and
- prestressed concrete.

5.5.1 Steel

As materials improve, techniques become more proficient and cheaper. Four structural forms are utilised:

- beam;
- arch;
- suspension; and
- cantilever.

The simplest form of bridge is the **beam**. The shape can be varied along the length of the span, by increasing its depth gradually, close to the supports.

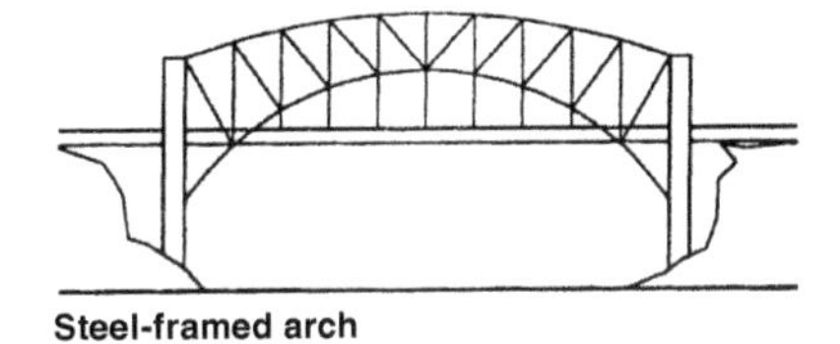

Steel-framed arch

Cantilever

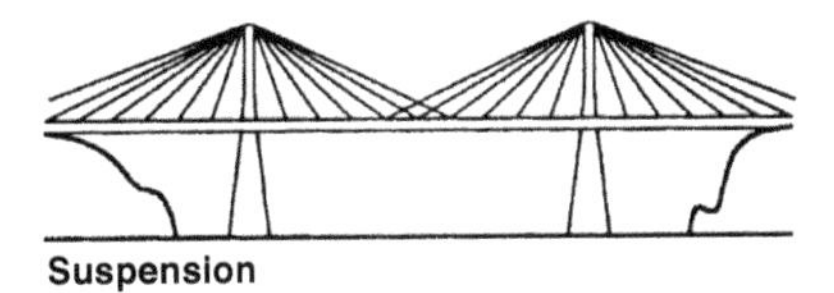

Suspension

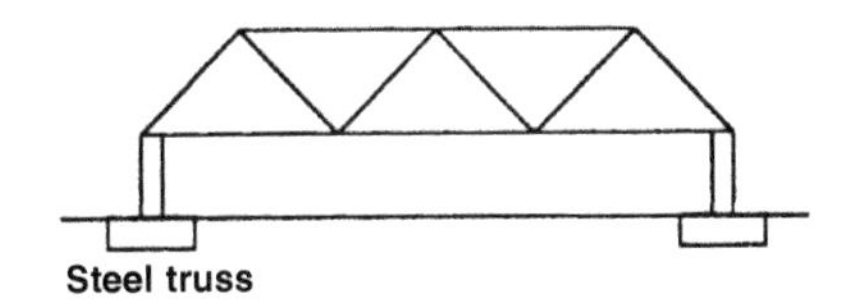

Steel truss

Fig 5.15 Bridge types

A continuous beam extends over several spans and the piers become props holding up one long beam. These methods can be used if an increase in the length of span is required.

An **arch** bridge has an upward convex shape, whereas a **suspension** bridge is exactly the reverse shape (concave upwards). The stress relationship is also opposite.

Cantilever bridges are the closest rivals of suspension bridges (in length of span). Due to technological advancement and improved construction techniques, these bridges can now span 50–150 m.

Activity 5.1

Select one type of bridge in this unit and find out as much as you can about it. Use the internet and the library to source information.

A new phase of bridge building was introduced when the world price of steel dropped by 75% in the 1870s. The new material was more versatile in bridge construction and bridging methods reached new peaks of development. Steel provided its own possibilities of new forms and enabled concrete to evolve as an effective medium. These changes contributed to the erection of many steel bridges. There have been three 'phases' in steel bridge construction.

5.5.2 Concrete

As a material for bridge building, concrete had a less immediate and smaller impact than steel. Today, concrete is used in an array of bridge building projects and is the chief rival of steel because of its relative cheapness. Concrete is a 'conglomerate' of strong, durable aggregates, bound together by a matrix of cement paste. Cement hardens and gains strength over a period of time. If steel reinforcement is laid within the concrete mass it increases the tensile strength of the concrete. Concrete can either be cast *in situ*, or be pre-fabricated (precast) in factories.

Whenever concrete is used, allowances must be made for shrinkage and creep.

The first concrete bridges

In 1840, in France, a concrete bridge was built over the Garonne Canal at Grisoles. In the early use of concrete for bridges, only its great compressive strength was exploited. Until the 1890s, mass concrete was used in bridge superstructures and foundations (mass concrete has no reinforcement and comprises only concrete). In 1871, in the United

States, a 10 m concrete arch was built in Brooklyn's Prospect Park to look as similar to stone as possible.

5.6 Reinforced concrete vs structural steel

When building a bridge or any other structure, one has a selection of materials to choose from, such as concrete or structural steel. Many people are not in favour of using structural steel because it rusts, while those who are in favour say that concrete, when overstressed, crumbles.

Choosing between concrete and steel for a bridge involves several factors including:

- span;
- foundation conditions;
- loads;
- architectural considerations; and
- economic considerations.

These two materials have excellent characteristics and produce the best results when combined. For example, reinforced concrete can be used for foundations and structural steel can be used for the superstructure. A further option is precast concrete. When building over a canal, river or railway, precast construction must be considered.

The advantages of one material appear to compensate for the disadvantages of the other. The great shortcoming of concrete is its lack of tensile strength (whereas this is one of the great advantages of steel). However, the two materials bond very well and can act together as a unit in resisting forces.

The excellent bond obtained is due to the chemical adhesion between the two materials, the natural roughness of the bars, and the closely spaced, rib-shaped deformations rolled on the bar surfaces.

Reinforcing bars are subject to corrosion, but the concrete surrounding them provides them with excellent protection.

Structural steel is composed almost entirely of iron with small quantities of other elements added to give it strength and ductility. Steel has high strength. Due to the range of sections available, steel is suitable for forming into shapes required for a particular structure. The speed of manufacture, transport and construction also make steel a desirable material for certain types of structural forms.

Structural steel members are made under controlled conditions which means that purchasers are assured of uniformity of quality. Standardisation of sections/shapes has helped to make design easy and keep costs down.

5.7 Prestressed concrete

More than 50 years ago, prestressing was almost unknown in bridge construction. One designer, Eugene Freyssinet, was dissatisfied with inadequate and highly theoretical approaches to bridge building and his contribution was the concept of prestressing.

In ordinary reinforced concrete, the cement and aggregate mixture is poured into forms or moulds, around reinforcing rods and bars. As the concrete hardens, the formwork bears the loading. When the forms are removed, the concrete, together with the reinforcing steel, is under stress. This transfer of loading combined with the expected shrinkage of the structural member, temperature fluctuations and possible vibration caused by construction works, can or will result in cracks in tension zones. The engineer must take this into account. One can cater for the self-weight of the structure given the accurate information available (dimensions, densities, etc.) and the stresses can then be accommodated during manufacture and construction by prestressing the concrete using steel cables.

5.7.1 Prestressed concrete beam

A deck slab can be constructed using prestressed beams, placed side by side on the bridge abutments and concreted as illustrated below.

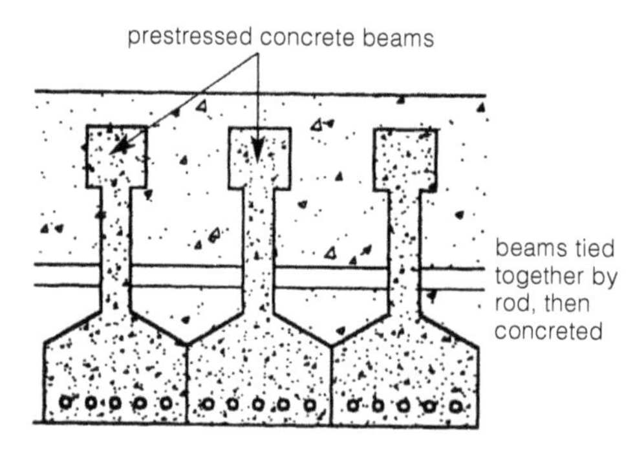

Fig 5.16 Transverse section of deck

This type of construction is needed for longer spans. The usual loading (prestressing) of the wires is about 1 000 N/mm^2. The wires are tensioned prior to concreting by means of jacking systems fitted to the beam mould (shuttering). The concrete is then poured. When hardened, the prestressed beam is removed from the mould. It tends to bend upwards (due to the prestressing) but deflects to an approximately horizontal position when loaded.

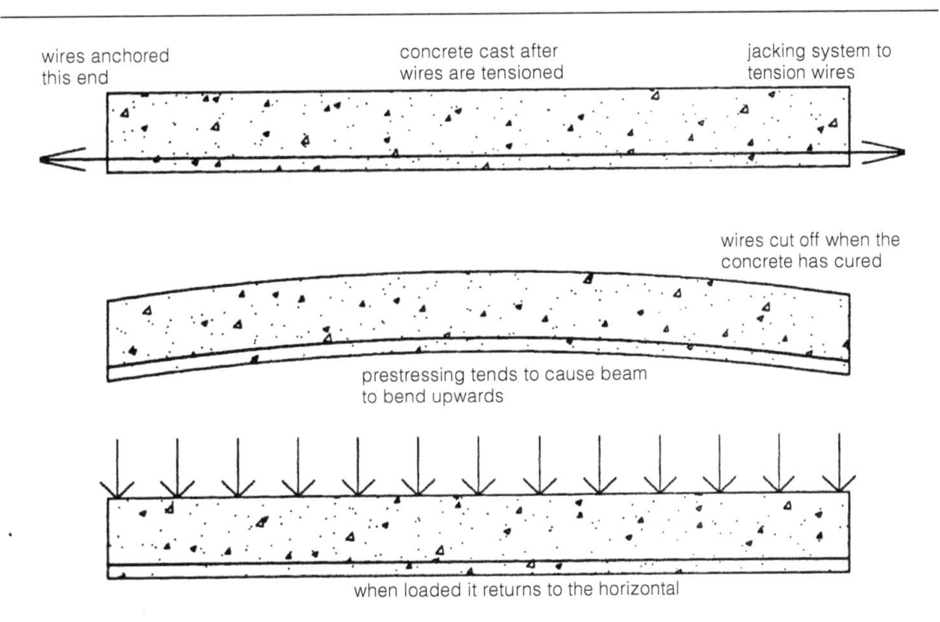

Fig 5.17 Prestressed concrete beams

5.7.2 Prestressing methods

There are two methods of prestressing that are still standard practices today:

Pre-tensioning

Steel wires in a mould are tensioned (fixed at one end and pulled from the other) and concrete is poured into the same mould. The wires are held in tension while the concrete gains strength and hardness. Once the concrete has gained sufficient strength, the steel ends are released, and the solid concrete holds them in tension. The wires, in turn, hold the concrete in compression.

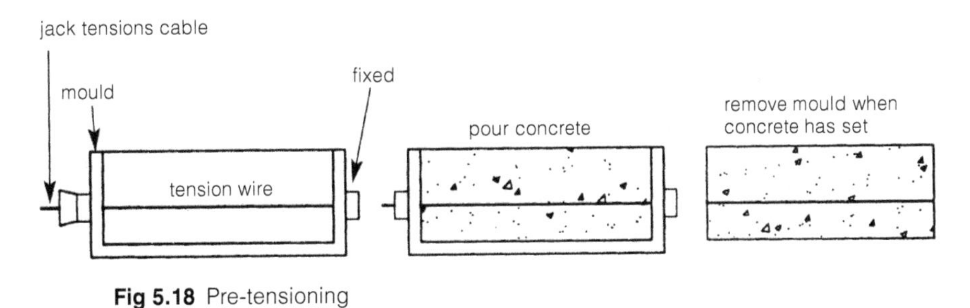

Fig 5.18 Pre-tensioning

Post-tensioning

Alternatively, the structural member can be cast without the steel but with holes or ducts running through it. When the concrete has cured, cables or bars are laid through the holes, stretched and then sealed in the beam with cement. This produces a post-tensioned member.

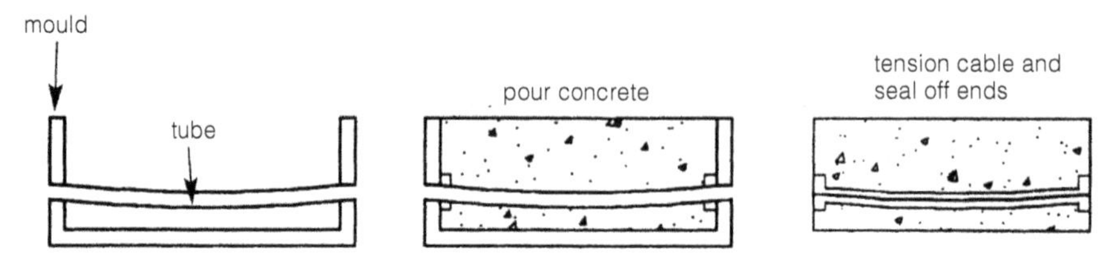

Fig 5.19 Post-tensioning

Table 5.1 Advantages and disadvantages of prestressing concrete

Advantages	Disadvantages
The implications of the techniques have been enormous and have led to prestressed concrete being used for short and medium spans in the construction of modern roads	Prestressed concrete requires higher-strength concrete and steel and a more complicated production process, resulting in higher labour costs.
Prestressing means that beams, slabs, etc. can be mass-produced under factory conditions and not necessarily cast on site. Quality can be easily controlled and the costs lowered. This is mainly due to the reduction in need for complex form-work, scaffolding, construction time and labour.	Close control is needed in manufacture.
Precast units can be joined on site by stringing them together with high tensile steel rods or cables to produce a complete prestressed structure. Such construction uses at least 70% less steel and between 30 and 40% less concrete than ordinary reinforced concrete.	Losses in initial prestressing forces, additional stress conditions that must be checked in design, and end anchorage and bearing plates all increase the cost.

Activity 5.2

Form groups of 4–5 and find information about and examples of the various types and forms of bridges discussed. Use this information to compile a report of not less than 10 pages and create a poster which can be used to make a brief presentation to the rest of the class.

Also, investigate the differences between steel and concrete bridges and tabulate your answers to identify advantages, disadvantages, ease of construction, costs, quality, strength of materials, size, etc. (e.g. if there were a choice between a concrete or steel bridge in any situation, what decision would the designer be faced with?). ➤

Also use the internet to determine the following information:

- The oldest bridge constructed and also the oldest bridge still in use today
- The bridge with the longest span in the world today
- What materials this bridge was made of
- How long it took to construct this bridge
- What method of construction was used
- What the challenges experienced during both design and construction were.

5.8 Foundations

Many factors influence the choice of foundation for bridges. The main factors will be the type and depth of the subsoil, and cost.

If an investigation finds that firm ground is available beneath deposits of soft clay, but that it will be uneconomical to excavate for a solid foundation, a decision may be made to use piles. Piles are used to transfer the load from a structural member to a solid foundation that can be at a depth below the base of the structure.

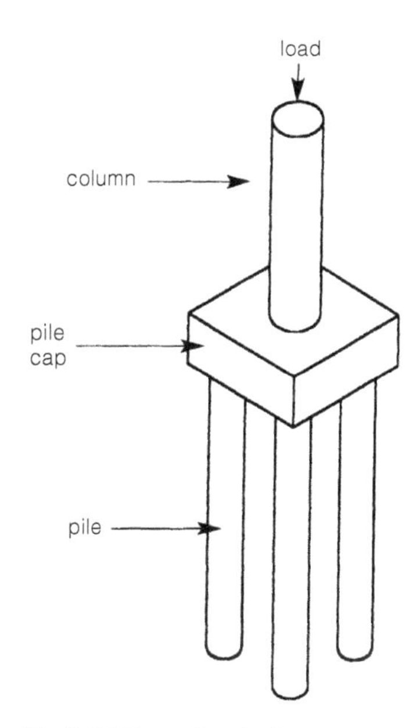

Fig 5.20 Four-pile cluster

The longest steel piles are those for the Tappan Zee Bridge. These piles are steel cylinders filled with concrete, with a maximum length of 82 m. The total length of all the piles used came to just over 10 km.

The longest bored piles were built for a river crossing of the Parana Guaze Bridge in Argentina. They are 70 m long, fully cased and carry a load of approximately 2 500 kg per pile.

Why do you think it was necessary to drive piles so deep?

When designing bridge structures, it is important that the engineer takes wind loading into account. Imagine placing a bridge in a valley that often experiences strong winds. Wind causes extra stresses that must be considered in bridge design.

Other examples of horizontal loading are, for live loads, the deceleration of traffic on a bridge, or the impact of ships against a quay or jetty. The greatest load in this category is that resisted at the anchorage of a suspension bridge. It is important that the anchorage remains fixed under load as this could cause structural instability.

A massive concrete pier that is keyed into bedrock is the favoured anchorage.

5.9 Summary

The purpose of this chapter was to:

- Identify different types of bridges, their functions and construction methods
- Explain the part each component plays in the complete structure
- Explain the criteria necessary in the design of bridges
- Discuss the various materials from which bridges can be made.

Self-evaluation 5

1. Compete the sentences:
   a. _______ bridges use two main cables spanning between towers.
   b. _______ are defined as being supports to the end span of a bridge deck.
   c. ___________ are used principally to accommodate horizontal expansion.
   d. The simplest form of bearing is a _______ _______ or _______.
   e. _______ is encountered when steel wires are tensioned before the concrete is poured into the mould.
2. State whether the following are true or false:
   a. The Golden Gate Bridge is an example of a cable-stayed bridge.
   b. Piers refer to the structures that support the deck.
   c. When using cables suspended between two towers, the cables are in tension.
   d. Concrete has high tensile strength.
   e. It is not important to consider the effects of wind loading when designing bridges.
3. Answer the following short questions:
   a. What is the difference between a suspension bridge and a cable-stayed bridge?
   b. How are the piers different to embankments in terms of support function/s?
   c. Describe the function of bridge bearings.
   d. List the properties that must be considered when designing a bridge.
   e. Describe your understanding of the difference between prestressed and post-stressed concrete.

Answers to self-evaluation 5

1. a. suspension
   b. abutments
   c. expansion joints
   d. rubber strip or pad
   e. prestressing
2. a. false
   b. true
   c. true
   d. false
   e. false
3. a. see refs 5.2.2, page 140 and 5.2.5, page 141
   b. see ref 5.3.2, page 143
   c. see ref 5.3.4, page 144
   d. see ref 5.4, page 146
   e. see ref 5.7, page 150

Chapter 6

Tunnels

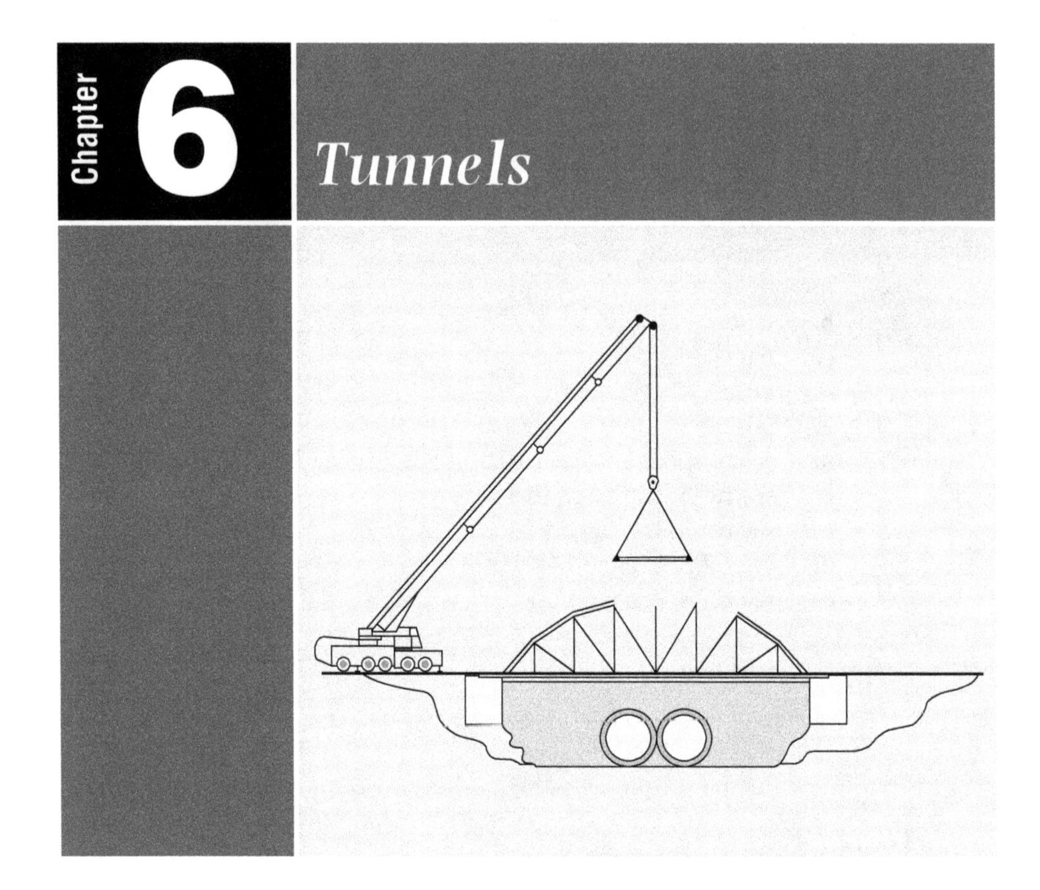

Outcomes

After studying this unit, you should be able to:

- Explain the purpose of tunnelling
- Discuss various tunnelling methods
- Compare the difference in application of tunnels in varying soil and rock conditions
- Identify some of the tunnelling machinery used in construction.

6.1 Introduction

One's first image of tunnels is usually those dug out by miners to move excavated material from the work face or to gain access to deeper sources. Tunnels are used for a variety of other reasons too, including road and rail transport, drainage, hydro conduits and other services. The most common types of tunnels are those used in roads and along railway lines. There are generally considered to be three types of tunnels:

- mining tunnels;
- public works tunnels; and
- transportation tunnels.

Mining tunnels are self-explanatory in that they are used to extract minerals and ore from the earth whilst mining. These tunnels are rather unstable and newsworthy because often you hear about accidents or collapses within these tunnels with dramatic effect (i.e. miners being trapped underground or sometimes killed).

Public works tunnels are mostly used for underground services like storm drains or large sewerage drains. Older and large cities like London and New York have large conduits (tunnels) under the city used to transport (convey) storm water or sewage away from the city.

Transportation tunnels used for road and rail services are generally the most expensive type of tunnels to construct.

Look around you and see how tunnels are incorporated into the environment. In South Africa, the most famous road tunnel is the Huguenot Tunnel in the Western Cape where approximately four kilometers have been dug through the Hottentots Holland range to provide road access between the Karoo and the coast. The 'Underground' rail tunnel system in London is another well-known example. A tunnel that has received much publicity is the one running beneath the English Channel, connecting England to the European mainland.

To read more about different types of tunnels, go to the websites http://science.howstuffworks.com/engineering/structural/tunnels1.htm as well as http://en.wikipedia.org/wiki/Huguenot_Tunnel

The methods used in tunnel construction vary from hand excavation (initially used in the mines) to more sophisticated power-driven tunnelling machines. The method of construction will depend on several factors, including:

- the ground type and conditions;
- the length of the tunnel;
- the expertise of the work contractor/work team;
- the type of tunnel required and its function;

- economics; and
- the stability of the rock.

6.2 Definitions

- A **heading** is the work face of a tunnel where blasting/excavation activities occur.
- **Lining** is a material (steel, concrete, precast concrete) used as an interior covering and to help strengthen the soil in the tunnel.
- **Pipe jacking** is a method used in the installation of pipes of varying diameters by forcing the pipes into the subsoil using a series of hydraulic jacks. Excavation is done while the driving of the pipe proceeds by hand or machine (depending on site conditions).
- **Full face** refers to the entire area of the heading where construction activities (blasting, drilling) take place.
- **Conventional shield** is a steel casing tube that is pressed into the heading in front of the lining to provide protection while the face is excavated.

6.3 Tunnelling

Six methods are used in tunnelling:

1. Drilling and blasting
2. Tunnel boring machines
3. Tunnelling with shields
4. Pipe jacking
5. Freezing method
6. Immersed tubes

6.3.1 Drilling and blasting

This method is mainly used when excavating in rock or where the diameter of the tunnel is too small for large excavating machines. Often short lengths of tunnel are excavated using this method. Traditional methods use explosives in excavations. Unlike in the past, when many of the blasts were uncontrolled, modern technology now makes it possible to be fairly accurate in achieving the desired effect. As mentioned previously, one of the factors affecting tunnelling is the stability of the rock. Several techniques are available to deal with varying situations:

- advance the heading without support;
- advance the heading in drifts followed by full-face support; and
- advance the heading in drifts followed by progressive support.

Full-face heading without support

This is used when the rock is fairly stable (after blasting) and can support itself without the need for propping. Usually tunnels with a cross-section of up to 200 m^2 are suitable. Holes are drilled, into which the explosives are placed, and then blasting takes place. The loose rock and debris is loaded into trucks and carted away. Once this is cleared, the operation continues until the tunnel is completed.

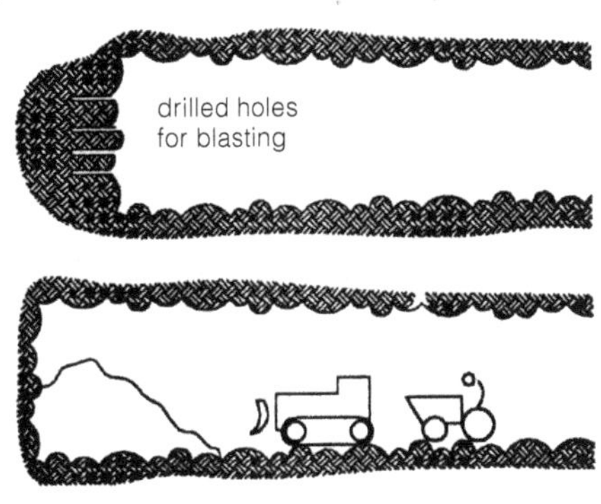

Fig 6.1 Full-face heading

Advance the heading in drifts followed by full-face support

The heading is opened in segments or drifts. If the rock is fairly stable and unlikely to suffer fracturing or bursting as the stresses in the rock are released, the drifts are sequentially advanced and temporarily supported until a complete ring of permanent lining can be put in place. As the heading is enlarged, props are used for initial support. The permanent lining is constructed when a full face has been excavated and the props are then removed. The face is progressively advanced in this manner.

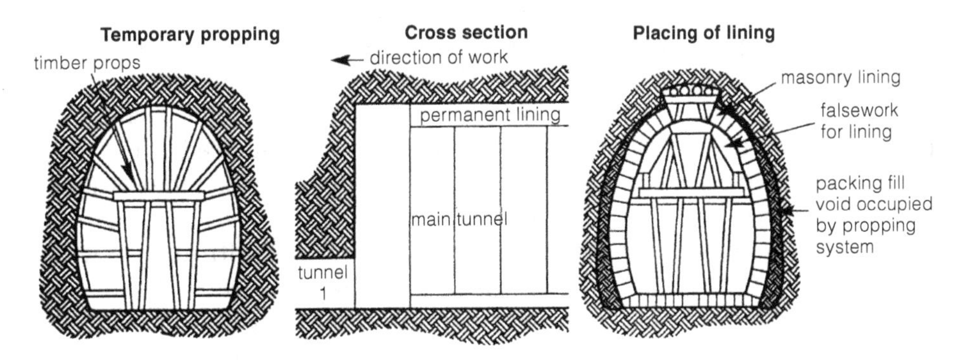

Fig 6.2 Excavation activities

Advance the heading in drifts followed by progressive support

This method is used when soft or fractured rock is encountered. It then becomes dangerous to leave the full face of the tunnel unsupported and permanent lining is installed progressively as the drifts are advanced. Different methods such as tunnelling with a shield are used for excavating and propping until the lining is in place.

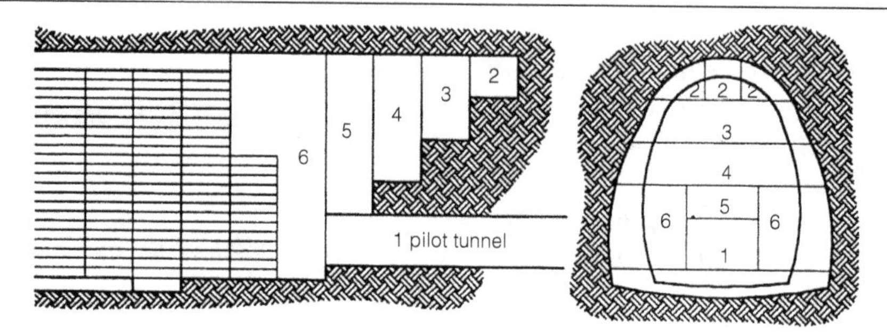

Fig 6.3 Typical excavation sequence

6.3.2 Tunnel boring machines (TBMs)

This method uses fairly sophisticated machinery to excavate and form tunnels. This is a direct response to technological development and the need for more economical tunnclling methods. This technique is employed in longer tunnels, where the cost of using traditional methods becomes too expensive. Like all machinery, it has its limitations, especially if the nature of the ground is likely to change frequently – for example, from soft ground to hard rock. This variation may render the cutting head unsuitable and it may be necessary to apply other techniques – for example, the shield method. The rotary cutting head of the tunnel boring machine requires a free-standing face on which to operate and therefore confined spaces or loose, falling rock may be problematic.

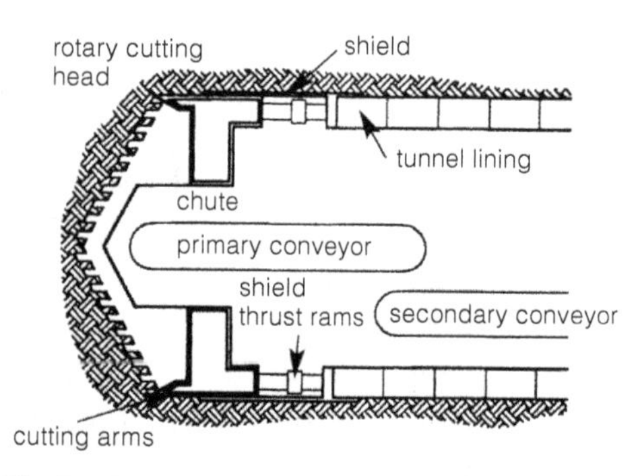

Fig 6.4 Tunnel boring machine

A tunnelling machine is sometimes combined with a conventional shield. The cutting head consists of three to eight radial arms fitted with chisels or discs which are rotated at up to eight cycles per minute. Excavated material is directed centrally through the cutting head and transported by conveyor to an area away from the working area.

6.3.3 Tunnelling with shields

In very soft, unsupported soil where the soil cannot be sustained long enough to place the permanent lining, the shield technique is applied. The shield acts as a steel casing tube which is pressed into the heading in front of the lining, thus providing protection while the face is excavated. The excavation of the face is usually carried out with power-assisted hand tools or mechanical excavating equipment. This method is suitable for tunnel diameters of between 4.00 and 12.00 m.

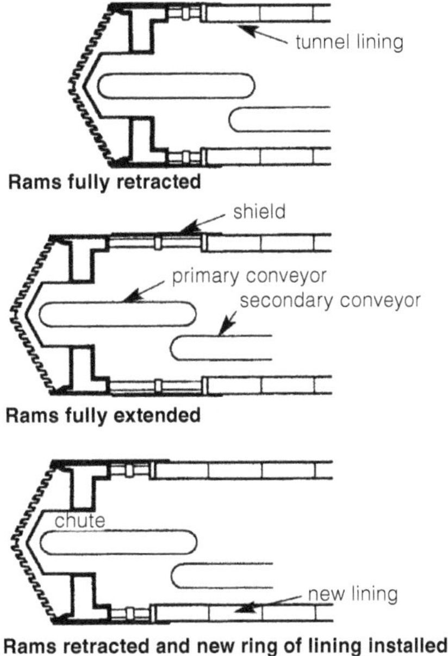

Fig 6.5 Tunnellng with shields

6.3.4 Pipe jacking

Developments in pipe jacking and auger boring are often referred to as mini-tunnelling for diameters of 0.75 to 4.00 m.

The demand for short lengths of medium-diameter tunnels for sewers and ducts in a single drive of up to 500 m is increasingly being satisfied using pipe-jacking methods. Soft rock, soft/medium soil and water-bearing granular materials can all be tackled using a system of concrete or steel pipes jacked (pushed) into position from a pit.

The jacking procedure involves pushing sections of pipe through the ground from a jacking pit using primary rams. Excavation takes place behind a shield as the driving proceeds.

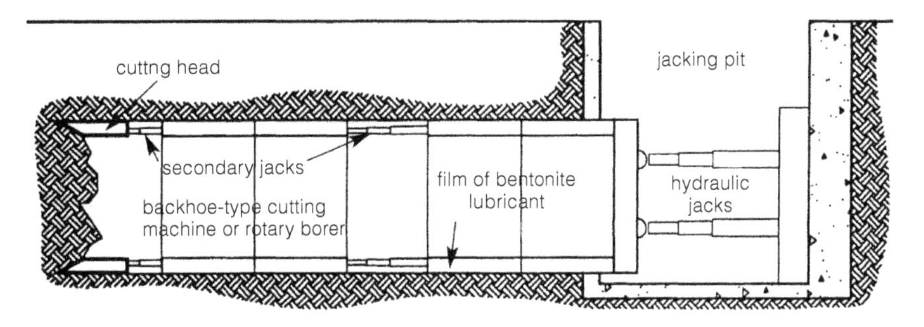

Fig 6.6 Pipe-jacking method

6.3.5 Freezing

Ground freezing is a method of construction that was used very successfully during excavations for the Huguenot Tunnel near Cape Town. Because the nature of the soil was very unstable, the only way to stabilise the soil condition was to introduce a brine solution that was piped into the soil. On contact, this would 'freeze' the ground in the vicinity of the pipes, allowing more stable excavation. Once the ground was 'frozen', more conventional tunnel-boring methods could be employed.

6.3.6 Immersed tubes

Immersed tubes are used for underwater tunnels. Precast units up to 100 m × 40 m are laid end-to-end to form the tunnel. The procedure consists of:

- dredging the trench in which the tunnel is to be placed;
- preparing either a permanent screed bed or piled foundation, depending on the soil, to support the tunnel sections;
- sinking the tunnel units; and
- sealing the joints.

Reinforced concrete or steel units are partially constructed on dry land, temporarily sealed at each end, launched into the water and built up to full dimension, then towed into position. The final position is checked by divers and adjusted if necessary. The joints between the segments are initially sealed with temporary gaskets and subsequently sealed with concrete or welded in place by divers. To reduce the risk of movement, the space between the hard core bed and the underside of the unit should be filled with coarse sand.

6.3.7 Modern non-tunnel boring machine (non-TBM) methods

Because of the high cost of timbering and the specialist skills required to install the initial supports, the traditional methods of tunnelling have gradually been modified as technical improvements to support systems and permanent linings have been developed. Modern non-TBM methods use one or more of the following:

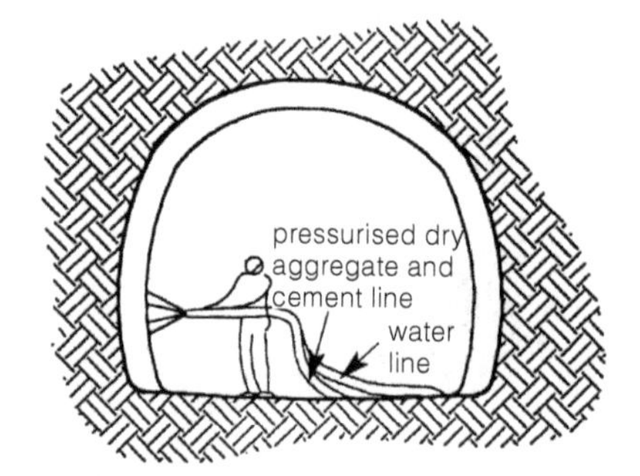

Fig 6.7 Sprayed concrete lining

- **Sprayed concrete.** The use of sprayed concrete of aggregate size up to 25 mm to which chemical strengthening improves adhesion and setting times.
- **Rock bolting.** The use of steel bolts in conjunction with sprayed concrete to provide support to tunnel roofs. The number of bolts and the spacing required to support a given area will depend on the diameter, material and type of anchorage of the bolt.

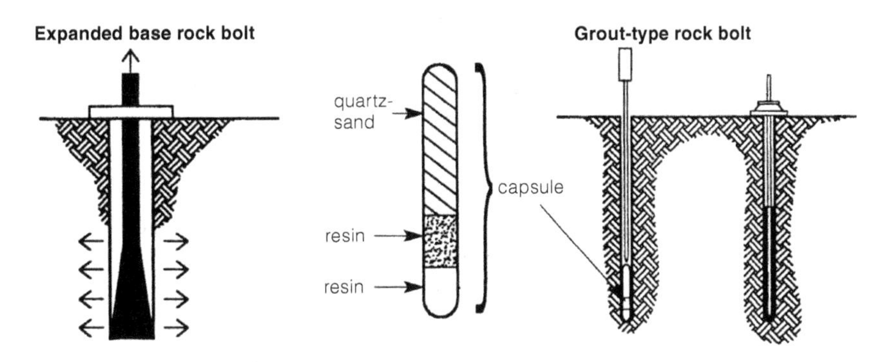

Fig 6.8 Rock bolting

- **Stiffening ribs and liner plates.** Where friable, shattered rock and certain categories of soft ground exist, and where immediate heavy ground pressure is encountered, neither spraying nor rock bolting is likely to be a suitable means of providing temporary support. In such conditions, additional propping with steel ribs is required. The face should be excavated in segments and parts of the rib temporarily propped until a complete unit of support can be formed.

- **Full lining.** When the permanent lining is designed in either cast iron or precast concrete, as is occasionally the case, it can then perform a similar function to liner plates. The usual method is to build a ring of linings from the bottom upwards. This method may be used in conjunction with rock bolting to increase working space, so that lining and excavation areas do not become too congested.

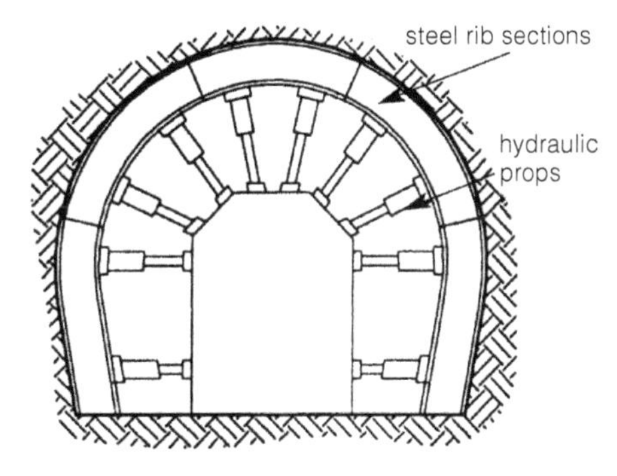

Fig 6.9 Full lining

6.3.8 Support systems and linings

When constructing a tunnel, it usually requires permanent lining which can also be utilised as temporary support during construction. This lining follows shortly after the excavation of the face of the tunnel. There are three common forms of lining used in tunnel design:

1. **Segmental forms** consist of cast iron units or precast concrete segments. The cast iron units are approximately 1.0–2.0 m long by 0.5–1.0 m wide and are bolted together to provide a strong

lining that can resist the external loads of the surrounding ground. Precast units were introduced to avoid the high cost of using cast iron. Joints are made to fit into each other and are covered with bitumen strips to prevent the ingress of water.

2. ***In situ* reinforced concrete**. Expensive cast iron lining in rock tunnelling is not required (in most cases the roof can remain unsupported) and reinforced sprayed concrete is adequate. If more strength is required, normal reinforced concrete can be cast behind specially designed travelling formwork.
3. **Masonry** is one of the 'older' tunnelling methods whereby the complete lining was constructed of masonry brickwork. Today these linings are mostly used to provide a protective coating for other lining systems in tunnels.

6.4 Summary

The purpose of this chapter was to:

- Identify the purpose of tunnelling
- Discuss the various types of tunnel construction methods
- Explain the role each part plays in the complete structure
- Discuss tunnelling in various ground conditions.

Self-evaluation 6

1. Complete the sentences:
   a. A ______ is the workface of a tunnel.
   b. ______ ______ is often referred to as mini-tunnelling.
   c. ______ ______ is the use of steel bolts in conjunction with sprayed concrete.
   d. Immersed tubes are used for ______ tunnels.
2. State whether the following are true or false:
   a. The nature of the ground type and conditions will affect the method of construction.
   b. Full-face heading is used where rock is unstable.
   c. Tunnelling with shields is used in very soft, unsupportive soil.
   d. TBMs are traditional means of excavating tunnels.
3. Answer the following short questions:
   a. What is a tunnel?
   b. Why do we use tunnels?
   c. Why may it sometimes be necessary to use tunnelling shields?
   d. Name the methods used in modern non-TBM tunnelling.
   e. Name and describe the three types of tunneling generally recognised as the basis of all tunnels.

Answers to self-evaluation 6

1. a. heading
   b. pipe jacking
   c. rock bolts
   d. underwater
2. a. true
   b. false
   c. true
   d. false
3. a. See ref 6.1, page 157
   b. See ref 6.1, page 157
   c. See ref 6.3.3, page 160
   d. See ref 6.3.7, page 162
   e. See ref 6.1, page 157

Chapter 7

Harbours

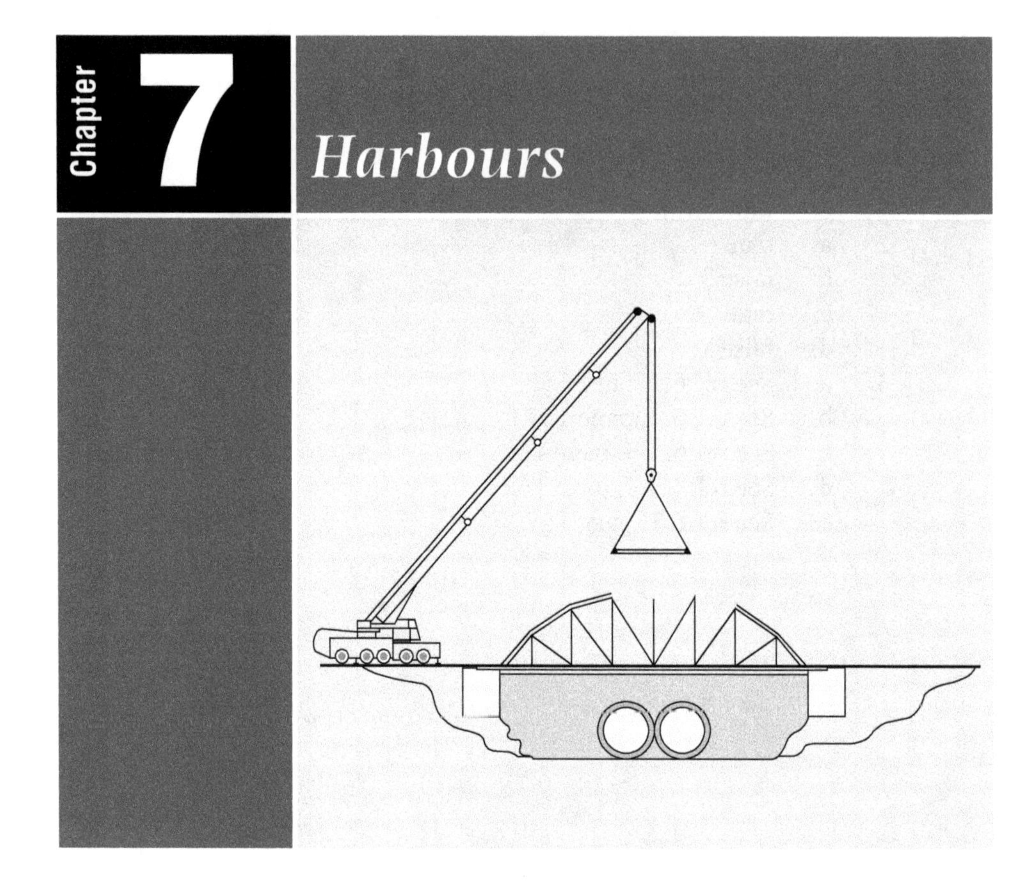

Outcomes

After studying this unit, you should be able to:

- Explain the functions of harbours
- Define the more familiar terms used in harbour design
- Discuss the various types of harbours
- Discuss and identify some of the structures and facilities of harbours.

7.1 Introduction

One of the more challenging projects an engineer can be faced with is the siting and design of a harbour. Given the unpredictable nature of the elements – wind, storms, tides, etc. – and the destructive force of water, careful planning and design is necessary to achieve a structure that will be suitable for its intended purpose. Harbours are built for many purposes, ranging from large commercial enterprises to small craft and recreational facilities.

You may have seen a large harbour like Durban, Richards Bay, Port Elizabeth or Cape Town, where various activities take place, including container handling, passenger embarkation, and the loading and offloading of commodities. You may have seen smaller harbours along the South African coast that cater mainly for fishing and support the activities of local residents. Whereas large and small harbours may follow similar design principles and planning, they serve different functions.

The following factors need to be considered in harbour planning:

- location and current condition;
- the need for harbour protection (e.g. against wave action);
- advantages and disadvantages of the harbour;
- keeping the project within reasonable limits of demand, finance and economic conditions;
- sources of financing;
- the type of harbour to be developed and its function; and
- the number, size and type of craft to be accommodated.

7.2 Definitions

- A **harbour** is an area where ships, boats and barges can seek shelter from the sea or in rivers, to dock and transfer people or cargo to land.

What is the difference between a 'ship', 'barge' and a 'boat'?

- **Wharf.** A platform made of timber, stone or concrete where the loading and unloading of ships takes place.
- **Dock.** An area between two wharves or piers used for mooring ships.
- **Dry dock.** An area of water that can accommodate a ship and be closed off. The water is then drained, leaving the dock dry, making it easier for repairs to take place.
- **Pier.** Similar to a wharf, a pier is a structure with a deck built out over water that is used as a landing place for ships.
- **Mooring.** A place used to anchor a ship.

- **Berth**. A place assigned to a ship at mooring.
- **Jetty**. A structure built from the shore that extends into the water.
- **Quay**. A wharf built parallel to the shoreline.
- **Breakwater**. A massive wall built out into the sea, that is used to protect the harbour from wave action.
- **Slipway**. This is the sloping area (ramp) in a harbour from which a vessel is launched.
- **Buoy**. This is a distinctly shaped and usually orange-coloured float that is anchored on the seabed to serve as a warning or to designate moorings.

7.3 Functions of a harbour

The functions of a harbour can be divided into various categories. For example, a harbour can be used as a storm or emergency haven, as a commercial venture, a recreational centre, or a fishing boat moorage.

The basic functions of a harbour are as follows:

- A harbour **shelters** vessels from the destructive forces of wind, currents and waves.
- Harbours can be used as navigational aids to **support ships at sea** (by using, for example, lighthouses, buoys, fog horns, markers, etc.). In larger harbours along the South African coastline, guidance in and out of the harbour is provided by tugs, pilots and communication networks.
- Elements of **service provision** must be considered – for example, refueling, telephone and medical services, stock replenishment etc. Ship building and repair yards also need to be considered. An efficient port maintenance organisation must exist to carry out repairs to structures within the harbour, as well as provide new structures. In many harbours it is essential to have a special unit to keep the harbour free of pollution. Dredging will need to take place if the harbour tends to silt up in order to prevent the waterways from becoming shallow.
- Harbours need to be able to **receive and dispatch cargo** via land and sea. These have to be transported by means of equipment such as cranes, forklifts and pipelines onto wharves and jetties, and then stored in areas such as warehouses, silos, container yards and bunkers, before being sent on to their final destinations.

7.4 Types of harbours

The function of a harbour usually determines the type of harbour that is designed. Most harbours are built to accommodate commercial activities such as ship building, the handling of goods and people, and commercial fishing. Others are built with a specific purpose. For example, Simon's Town harbour was constructed as a naval base.

One of the most famous historic war battles occurred in a place called Pearl Harbour on 7 December 1941, when American battle ships stationed there were bombed by Japanese fighter pilots. There was also an epic movie of the same name made in 2001 to dramatise some of the events that took place during this time. If you are intererested in reading more about this battle, you can check out this website: http://history1900s.about.com/od/worldwarii/a/Attack-Pearl-Harbor.htm

The Port of Southhampton, is a port well known for its ship-building exploits, having been responsible for launching many famous ships over the past decades, one of which was the infamous Titanic.

Some harbours have developed as **natural harbours** or shelters carved out of the land. Examples include New York, Rio de Janeiro, Sydney and Hong Kong. Some small vessel harbours have become **improved harbours** by providing breakwaters or removing sandbars. Examples of improved harbours include those in Durban and Venice. **Artificial harbours** are built on dry land and then opened to the sea. Examples of this type of harbour include those at Marseilles, Port Said and Cape Town.

7.5 Harbour structures and facilities

A typical harbour layout and the terminology used for various structures are given below:

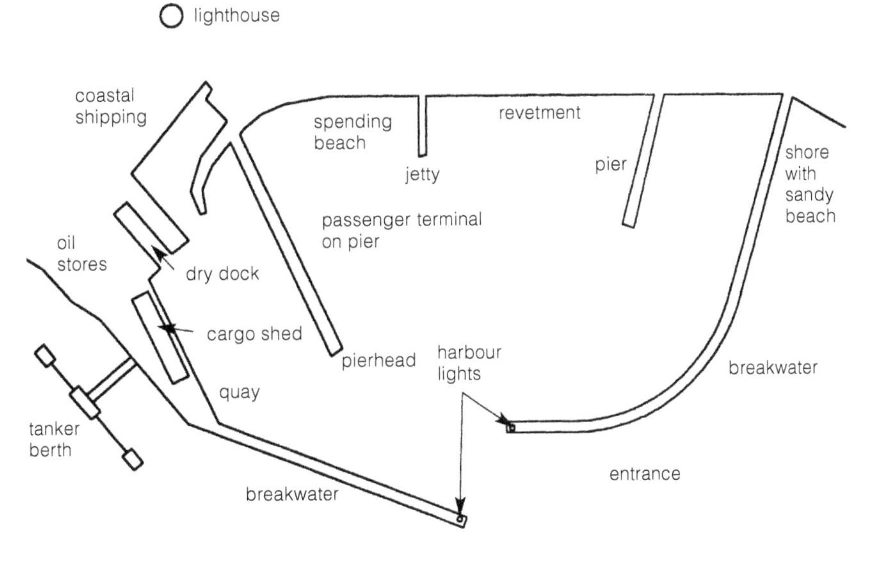

Fig 7.1 Typical harbour layout

7.5.1 Wharf

A wharf is a structure built on the shore of a river, canal or bay that makes provision for vessels to lie alongside it for the receiving and discharging of cargo and passengers. A **quay** (also called a **marginal wharf**) is built parallel to the shoreline. A wharf built at an angle to the shoreline is called a **pier** or **finger pier**. The berthing and manoeuvring space between adjacent piers is called a **slip**.

When deciding on the type of wharf to construct, it is important that the local conditions of the harbour receive priority. For example, an area that has limited waterfront land but has ample space for manoeuvring will be more suited to the construction of piers. In cases where there is ample waterfront land available, it may be better to construct quays. However, consideration must be given to the amount of dredging required to maintain berthing.

Wharf substructures

There are two substructure variations: **solid fill type** and **open type**. The solid fill type consists of a vertical gravity wall constructed on the waterfront face. Earth is then backfilled behind the wall and covered with a paving deck that is usually made of concrete.

In open type construction, the wharf superstructure is supported by timber, concrete or steel piles. Transverse rows of bearing piles are driven and capped with concrete girders. Decking is then placed on top of these girders.

7.5.2 Bulkheads, seawalls and revetments

A **bulkhead** is a vertical wall constructed of steel, timber or concrete piling and placed parallel or nearly parallel to the shoreline. It serves as a secondary line of defence in major storms. It also separates land areas from water areas.

The main purposes of a bulkhead are to retain or prevent landslides and to provide protection against wave action. A bulkhead should not be seen as a long-term solution, especially where it is exposed to the severe weathering. It may be necessary to enlarge the bulkhead into a massive seawall that is capable of withstanding direct wave action.

Seawalls can have vertical, curved or stepped faces. It is important to protect them against scouring and undermining, which is caused by the downward force of water when waves strike the wall. A stone apron is often necessary to prevent this happening. Apart from protecting the land from the scouring action of waves, a seawall also acts as a retaining wall. Seawalls only offer protection to the land immediately behind them and not to adjacent areas along the coast.

A **revetment** comprises one or more layers of stone or concrete on the sloping face of a seawall. The sloping protection dissipates wave energy, reducing the damaging effect of waves striking a vertical wall.

When a wave strikes the wall, water is forced both upwards and downwards. The downward movement of water is more hazardous to the stability of the wall as it tends to scour and erode the base (toe) of the wall. It is therefore important to have adequate and sufficient protection against this action.

Other shapes used to minimise the force are either concave or convex. A convex curved face and smooth slopes effectively dissipate and absorb wave energy, and reduce wave overtopping and run-up. Sloping face walls also reduce scoring when compared to vertical faces.

Concave curved face structures are effective in reducing wave overtopping and spray when onshore winds are light.

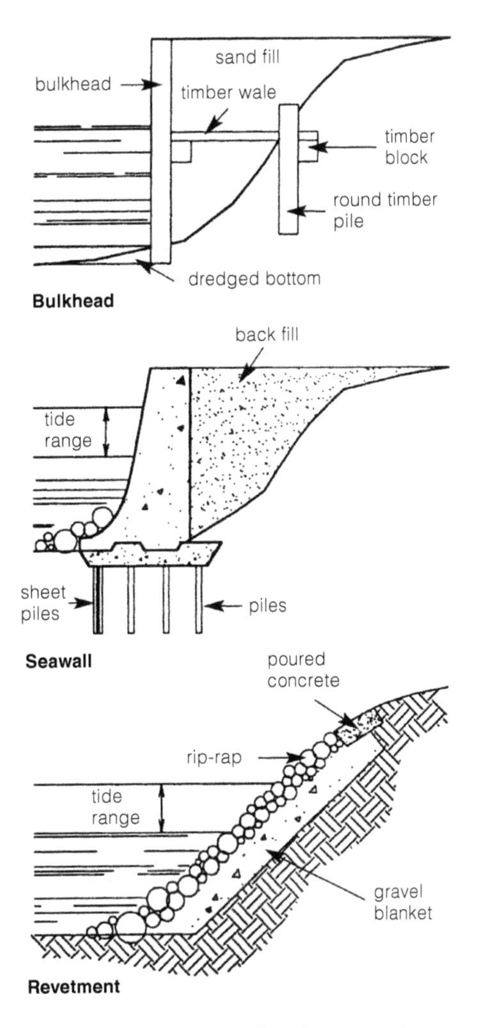

Fig 7.2 Bulkhead, seawall and revetment

It may be difficult to distinguish between a seawall, a bulkhead and a revetment as these features are determined at the functional planning stage. Generally, seawalls are the largest of the three because they resist the full force of the waves. Bulkheads are smaller as they mainly retain fill and are not exposed to the severe and direct action of the waves. Revetments are the lightest because they are designed to protect shorelines against erosion by currents and light wave action.

7.5.3 Jetties

A jetty is a structure that extends into the water to direct and confine river or tidal flow in a channel. It is also used to prevent or reduce silting of the channel. Jetties located at the entrance to a bay or a river also protect the entrance channel from wave action and cross current.

Jetties are usually constructed of steel, concrete or rock, and can be built on steel sheet-pile cells, caissons, or a single row of braced and tied timber piling. Choosing the most suitable type of jetty will be influenced by the foundation conditions, wave action and economic considerations.

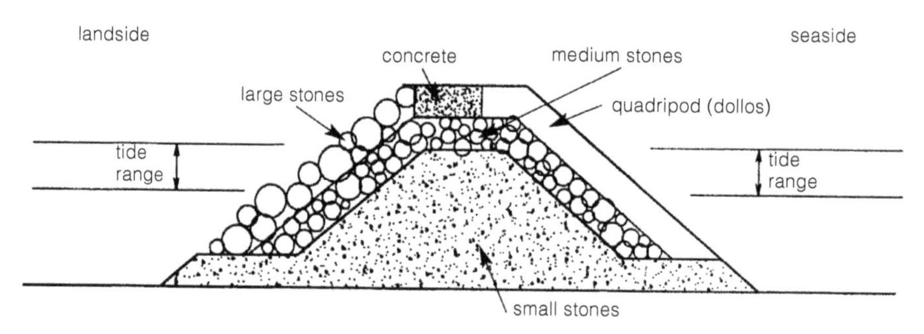

Fig 7.3 Quadripod-rubble-mound jetty/breakwater

7.5.4 Breakwater

A breakwater is a structure built to protect a harbour, shoreline or basin against wave action. A breakwater for a harbour results in calmer water inside the enclosed area, whereas a breakwater protecting a shoreline results in reduced wave energy reaching the beach. Other functions of a breakwater structure include providing protection for safe mooring, the operating and handling of ships, and harbour facilities.

There are two broad categories of breakwater: offshore structures and onshore (shore-connected) structures. Offshore structures are constructed mainly for navigation purposes or to shelter a harbour entrance, and are more costly than onshore breakwaters.

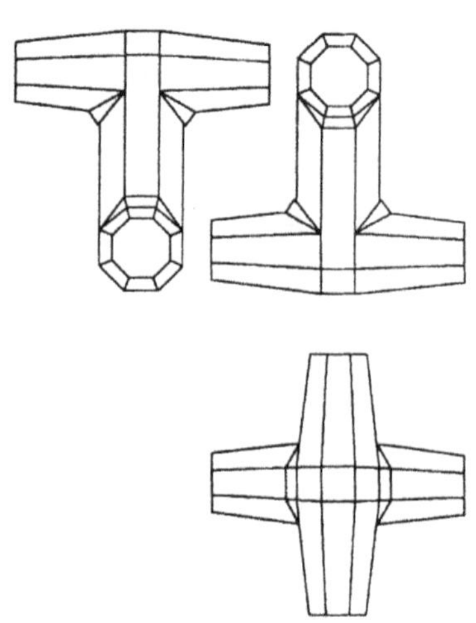

Fig 7. 4 Dollos

There are three main types of breakwater:

- vertical sides;
- rubble mounds; and

- composite structures (where a rubble mound is used as the major foundation and carries a vertical-sided superstructure).

In South Africa, it has become common practice to replace rubble mound breakwaters with 'dollos'. Dollos are designed concrete sections that have interlocking properties which allow them to knit together closely once placed. The shape of a dollos is also specially designed to counteract the destructive forces of waves.

Activity 7.1

Next time you visit a harbour, see if you can identify some of the elements discussed in this unit.

What is the difference between a 'port' and a 'harbour'?

Read more about harbours and ports on the following websites:

- http://www.southafrica.com/ports/
- http://en.wikipedia.org/wiki/Ports_and_harbours_in_South_Africa
- http://www.engineeringnews.co.za/page/harbour-infrastructure-and-equipment

7.6 Summary

The purpose of this chapter was to:

- Explain the functions of harbours
- Define the more familiar terms used in harbour design
- Discuss the various types of harbours
- Discuss and identify some harbour structures and facilities.

Self-evaluation 7

1. Complete the sentences:
   a. A structure with a deck built over the water used for mooring ships is called a _______ .
   b. A _______ shelters vessels from the destructive forces of wind, current and waves.
   c. A _______ is a vertical wall constructed of steel, timber or concrete piling and placed parallel to the shoreline.
   d. A breakwater is a structure built to protect a harbour against _____________.
   e. _______ are designed concrete sections that have interlocking properties and are used to protect the shoreline.

➤

2. State whether the following are true or false:
   a. A slipway is an area in a dock where repairs take place.
   b. A jetty is a structure built from the shore that extends into the water.
   c. Seawalls comprise one or more layers of stone or concrete.
   d. There are three categories of breakwater.
   e. Dollos are used to dissipate wave energy.
3. Answer the following short questions:
   a. Describe the basic functions of a harbour.
   b. What is the difference between a wharf and a pier?
   c. Can one make distinctions between bulkheads, seawalls and revetments and, if so, what are they?
   d. Name the three types of breakwater.

Answers to self-evaluation 7

1. a. pier
   b. harbour
   c. bulkhead
   d. wave action
   e. dollos
2. a. false
   b true
   c. false
   d. false
   e. true
3. a. See ref 7.3, page 168
   b. See ref 7.5.1, page 170
   c. See ref 7.5.2, page 170
   d. See ref 7.5.4, page 172

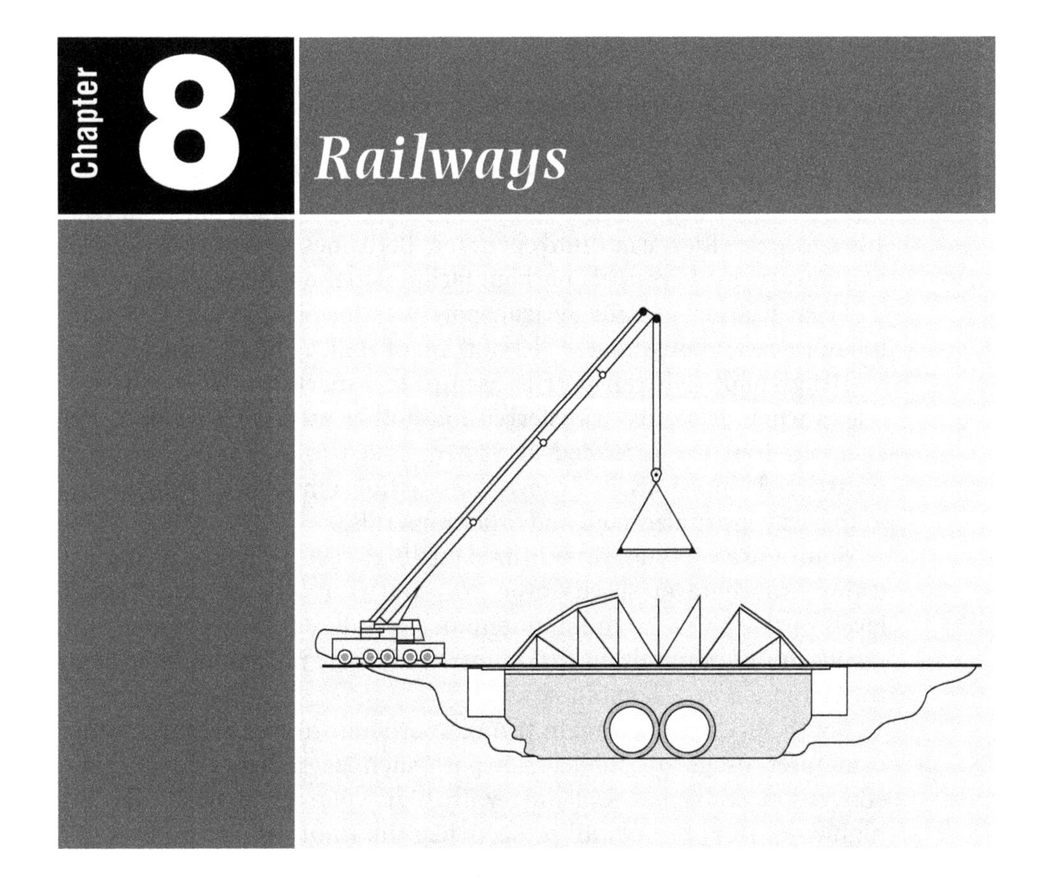

Chapter 8 Railways

Outcomes

After studying this unit, you should be able to:

- Understand the function of a track structure
- Define the terms used in railways
- Identify and discuss the various terms of a track structure
- Discuss the function of each element of a track structure
- Discuss the maintenance of railway lines.

8.1 Introduction

Because South Africa lacks navigable rivers but has long haulage distances, railways offer the ideal means of mass transport. The first railway line in South Africa was opened in 1860 and consisted of three kilometers of track between Durban and Point. Two years later, a 34 km line from Cape Town to Eerste River came into operation. Both lines were privately owned and constructed to connect harbour facilities to the nearest towns.

Electrification of railway transport was introduced in 1925 and, today, there is approximately 40 000 km of railway line in operation.

Heavy-haul rail transport is used to transport ores from mines to places where it can be transferred to another mode of transport. For example, iron ore is loaded at Sishen, railed to Saldanha and then transferred onto ships for export. Among the commodities transported in this way are maize, coal and other minerals.

South Africa has not yet ventured into high-speed intercity passenger trains (travelling at speeds over 200 km/h) as this requires higher levels of track design and maintenance standards. Unfortunately, the popularity of commuting (passenger transport) by rail declined with the introduction of kombi-taxis.

South Africa is once again trying to promote and encourage higher commuter usage of public transport such as railways to alleviate congestion problems associated with high volumes of traffic on our highways. High profile rail projects like the Gautrain are a means of addressing this problem for commuters between Pretoria, Johannesburg and OR Tambo International Airport.

8.2 The permanent way

The permanent way or track structure is the most important fixed asset of a rail transport system. Like all other engineering disciplines, it is important that correct planning and design considerations are taken into account when establishing a railway line. Once construction is complete, regular maintenance forms an integral part of this transport system.

A rail system utilises several components including steel rails, timber or concrete sleepers, and stone ballast. Each component responds in a particular manner when exposed to different conditions. Together with this are the dynamic forces imposed by the train and its load. It is therefore important that the function and construction methods employed in each element are understood, so as to ensure safe and convenient travel.

8.2.1 The function of a permanent way

What do you think the function of the track structure is? If you are thinking about guidance, support and stability, you have the right idea. The function of a permanent way is to provide lateral (side) guidance and support to the rolling stock (rail trucks) using the railway line.

Have you noticed that train wheels are grooved? The wheels fit over the steel rails, which are supported both laterally and longitudinally on sleepers spaced at uniform (equal) distances. Steel fasteners ensure that the rail does not move from the sleeper. Underneath the sleepers are large stones called **ballast** to provide further support and stability. All these elements rest on a prepared formation.

8.3 Definitions

Figure 8.1 indicates a section through the track structure and defines the most commonly used terms.

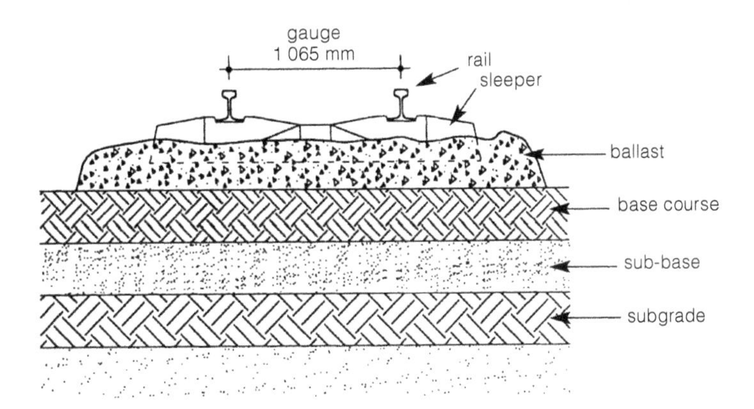

Fig 8.1 Cross-section of a typical formation

- **Rail.** This is made of steel and used as a guide for the rolling stock.
- **Sleeper.** Recently made of concrete and used to support the rail.
- **Ballast.** Large aggregates (> 37 mm) overlaying the formation and supporting the sleeper/rail combination. It also provides protection for the formation.
- **Formation.** Similar to that of a road structure, consisting of compacted *in situ* or selected material.
- **Gauge.** The horizontal distance between two rails taken centre to centre. In South Africa, the gauge is 1 065 mm.

8.4 Rails

Think of the rail acting as a steel beam and remember the forces and movements that beams are subjected to (vertical and horizontal movements). What do you think can cause these movements?

Rails are subjected to a variety of forces in both vertical and horizontal directions – for example, from moving wheels, temperature changes and trains braking and accelerating.

Why do you think temperature is so important?

Because the rail is made of steel and steel expands and contracts in response to temperature changes, this will account for movement.

Rails are also subjected to shock loads at joints and at points where there are surface defects. Flat wheels – wheels that are not completely round – also add stresses to the rail.

All the rails used in South Africa are manufactured and supplied by ISCOR, except for the existing rails on the Sishen–Saldanha line which were imported from Germany.

Nine different rail profiles are used in South Africa:

- 20 kg/m
- 22 kg/m
- 25 kg/m
- 30 kg/m
- 40 kg/m
- 48 kg/m
- 57 kg/m
- UIC-60 kg/m (used for the Sishen–Saldanha line)
- S-60 kg/m

The 20, 22, 25 and 40 kg/m profiles have become obsolete. The rest are used as follows:

- 30 kg/m for narrow gauges of 610 mm;
- 48 kg/m for main and secondary lines;
- 57 kg/m for high speed and 20 t axle loads; and
- 60 kg/m for heavy haul (26 t axle loads).

8.4.1 Rail properties

Rails should have the following properties:

- They must be able to **withstand shock loads**.
- They must **resist high contact** and other **stresses** that cause abrasive forces and metal flow.

- They must **conform to close tolerances**, both in shape and chemical composition.
- They must **be chemically sound** so that they can be welded, cut, drilled, planed and shaped without causing undue defect to their composition.
- They must **be free from metallurgical and mechanical defects.**

8.4.2 Rail joints

There are three types of rail joints that have been classified as follows:

- welded;
- fish-plated; and
- insulated.

Welded joints

Rails are flash-butt welded after being placed on site using a Thermit welding method.

Fish-plated joints

This jointing method was very popular during the 1980s when a four-hole fishplate was used. Its popularity has slowly waned due to the increasing use of continuous-welded rail. Bolts and nuts are used to fasten the fishplate.

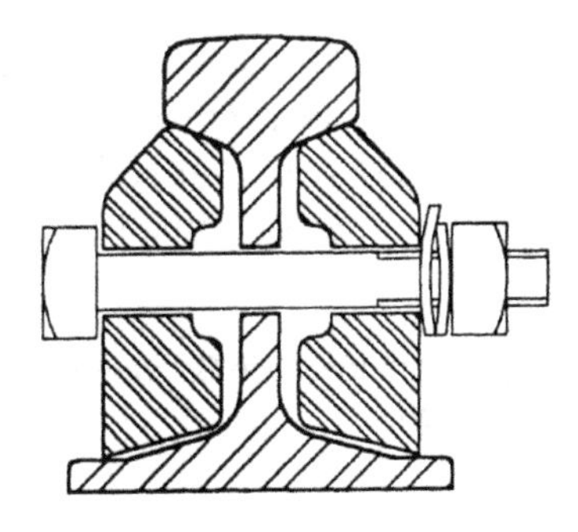

Fig 8.2 Fishplate connection

Insulated joints

These joints must be installed on lines with automatic signaling and centralised control. There are two types of insulated joints: those for a jointed track and those for a continuous welded rail (CWR).

Before 1982, a four-hole, laminated wooden fish plate was used for jointed tracks. This acted as the insulator, but because of its unsatisfactory performance, a glass-fibre fishplate superseded it.

Two types of insulated joints can be used for the continuous welded rail – a **four-hole glued** and a **six-hole dry joint**. In the first joint, the glue acts as a strengthener and an insulating agent. In the six-hole joint, an inert layer bonded onto the fishplate provides positive insulation, but no glue is used. Both the glue and six-hole joints are prefabricated in workshops, and then Thermit welded into position. Spoornet is investigating the use of electronic signalling devices that could reduce the use of insulated joints.

8.4.3 Continuous-welded rail

This type of construction is extremely popular given the reduction in the amount of track maintenance needed compared to conventional jointed track. The rail is prefabricated into lengths of 216 m, welded using continuous-welded joints and placed into position. The use of this type of construction is restricted by sleeper type, rail mass and track curvature.

8.5 Sleepers

The main purpose of sleepers is to support the rails in the correct geometrical position and to provide lateral and longitudinal stability to the complete track structure.

Three types of sleepers were initially used for track construction: wooden, steel and concrete. Given the high cost of steel and wooden sleepers (nearly twice that of concrete) and the technical advantages of concrete, it was decided to use concrete sleepers instead. Wooden sleepers are still used on sharp curves (< 200 m) and on open-top bridges as they provide extra lateral stability on these curves (where concrete sleepers were found to be unsuitable).

Nearly 50 million sleepers are in use in South Africa, of which 60% are concrete, 20% steel and 20% wooden.

Sleeper spacing depends mainly on the loads they will carry, but generally this will vary from 600 mm to 750 mm.

Other functions of sleepers include:

- spreading the load over a wider area of the ballast;
- maintaining the gauge width between the two rails; and
- providing the required inclined bedding (1 in 20) for the rails.

Can you guess how heavy one passenger carriage of a normal commuter train is?

8.5.1 Concrete sleepers

Concrete sleepers were first introduced into South African track construction during 1952. They are manufactured according to very strict specifications:

- Spoornet specifies dimensions and bending strength.
- Prestressed monolithic sleepers must be used.
- 28-day strength of 60 MPa is the required minimum.
- A 40-year service life is expected.

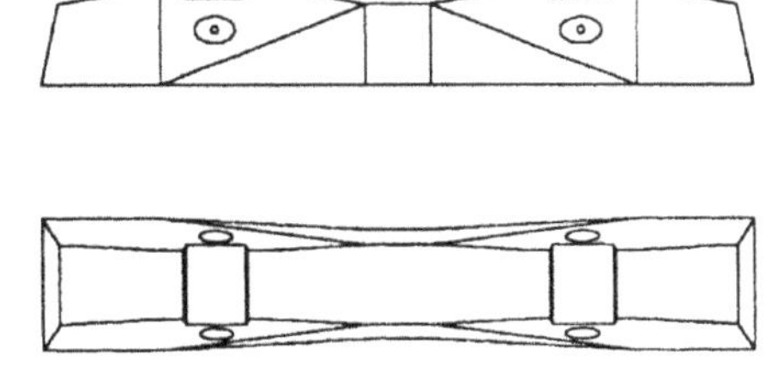

Fig 8.3 Concrete sleeper

8.5.2 Wooden sleepers

Wooden sleepers are generally made of hardwood. Indigenous hardwoods are unsuitable for this purpose, so they are imported from Mozambique, Australia and Zimbabwe. The standard dimensions of a wooden sleeper used on a 1 065 mm gauge line is 2 100 mm × 250 mm × 125 mm. Wooden sleepers are expected to last between 15 and 25 years, depending on rail traffic conditions. All sleepers are treated with a 70% creosote and 30% synthol waxy oil mixture.

8.5.3 Steel sleepers

Two types of steel sleepers were designed – one for 30 kg/m rails and the other for 40 kg/m rails. However, these rails were discontinued due to their high cost, lack of lateral resistance and difficulty in insulating for automatic signals.

8.6 Fastenings

Fasteners are used to hold the rails securely onto the sleepers and prevent undue movement. Initially, nuts and bolts or spring-clip fasteners were used to tie the rails onto the sleepers, until experiments with other fasteners showed improved results. It was then decided to standardise on two fastening systems: pandrol and fist fasteners.

Rail to sleeper fastening must:

- **provide a strong bond** between rail and sleeper, preventing movement and gauge loss;
- **prevent creep** of the rail longitudinally;
- **provide electrical insulation**, preventing short-circuiting of the current; and
- have **good fatigue resistance**.

8.6.1 Pandrol fastening

This uses a standard assembly consisting of a spring-clip, a glass-reinforced nylon insulator and a high-density polyethylene (HDPE) pad. To accommodate changes in the rail profile, insulators with different dimensions are used.

8.6.2 Fist fastening

The standard assembly consists of a spring-clip, a nylon-insulated pin and an HDPE plate.

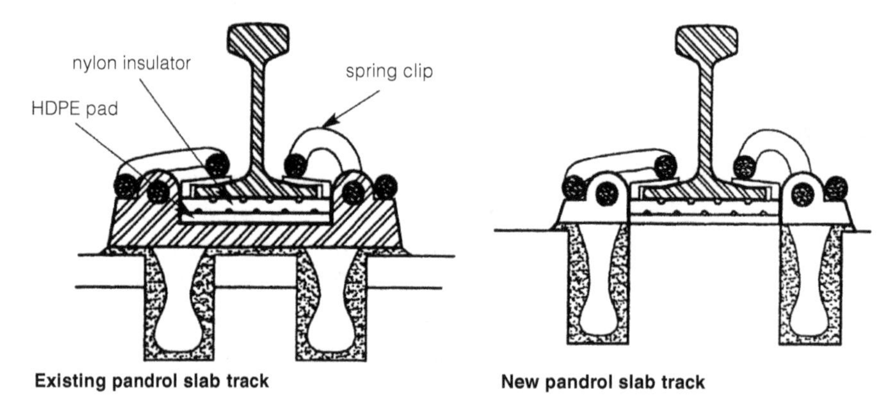

Fig 8.4 Pandrol slab tracks

8.6.3 E3131 rail to sleeper fastening

The standard fastening for wooden sleepers is the E3131 cast iron chair, which is equipped with T-bolts and clips and fastened onto the sleeper with coach screws.

8.7 Ballast

Ballast refers to the large aggregate used as a supporting medium for the rail and sleeper. The type of aggregate used is granite, dolomite, basalt, quartzite and tillite, but will vary from area to area based on availability and cost. Spoornet specifies that the aggregate must conform to tests for sulphate soundness, flakiness, durability and abrasion up to a maximum stone size of 63 mm.

The depth of the ballast may vary according to the circumstances and operation to which it is exposed. For example, for axle loads greater than 20 t on curves from 400 to 600 m radius, the depth would be 300 mm.

The quantity of ballast per kilometre of track will depend on the traffic density of the line (given in gross tons per annum). The dimensions of the profile will depend on the sleeper length and the quantity of ballast per kilometre.

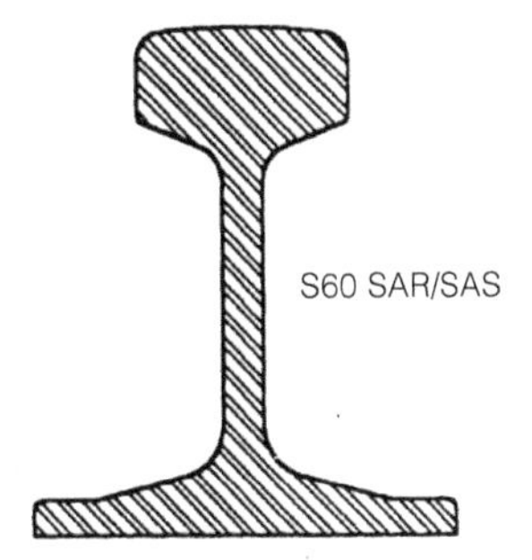

Fig 8.5 Typical rail profile

Annually, approximately 2.5 million cubic metres of ballast is used in South African rail systems. It is used to support the track in such a way that:

- the load is **distributed uniformly** over the formation;
- the sleepers are **prevented from lateral or longitudinal movement**;
- there is **good water drainage**;
- there is proper **air circulation**; and
- **efficient maintenance** of line, level and super-elevation can take place, either by hand or by mechanical means.

8.8 Formation

As in road construction, the formation is the foundation on which the track is constructed. It needs to maintain a stable track superstructure with minimum deviation over the design life of the track.

The formation absorbs and dissipates the dynamic loads of the rail trucks, but this load is partially absorbed by the ballast above it. The reduced load is then dissipated through shear stresses into the layered formation, which consists of imported selected material of a strength that decreases with depth.

The formation must be maintenance free because all the elements in the track structure would be affected (rail, sleepers, ballast, etc.). It is difficult to repair in the event of failure. To achieve a 'fail proof' formation, one needs to eliminate certain problems:

- **Drainage conditions** – when moisture gets into the formation it may lead to weakening of the soil material.
- **Changes in traffic loading** – ensure that any change in loading is compensated for by an increase in ballast support.
- **Construction materials** – ensure that suitable material, within specification, is used as local material may not conform to standard. (See also Fig 8.1 Cross-section of a typical formation.)

8.9 Turnouts and crossings

Turnouts are devices that allow a train to move from one track to a new, parallel track that starts at the end of the turnout area. A crossing is similar to a turnout and allows a train to switch from one track to another that is running parallel to it.

8.10 Grades

Have you tried moving metal on metal and noticed the low traction?

Trains do not rely on tractive forces between the wheels and the rail to supply motion. Instead, they rely on rolling motion. Because of the low frictional forces between the rail and the wheel, fairly flat grades are needed to prevent slippage. The ruling grade on all main lines in South Africa is 1:66 or 1.5%.

8.10.1 Grade requirements

- Grades should be as flat as possible.
- Large deviations in ruling grades between main lines and shunting areas should be prevented.
- Tracks used for staging trains may be on a grade of 1 in 400.

8.11 Track maintenance

One of the more important operations on a railway system is line maintenance. Today, this operation is highly mechanised, with specialist equipment performing operations that were previously done by hand. Some of the maintenance operations include ballast cleaning, track tamping and alignment, rail profiling and lubrication.

The following websites may be of interest to provide additional reading as well as historic and background information:

- http://inventors.about.com/library/inventors/blrailroad.htm
- http://en.wikipedia.org/wiki/History_of_rail_transport
- http://mysite.mweb.co.za/residents/grela/transnet.html
- http://www.sahistory.org.za/dated-event/first-railway-line-south-africa-between-durban-and-point-officially-opened
- http://www.railway-technical.com/track.shtml
- http://ptv.vic.gov.au/assets/PTV/PTV%20docs/VRIOGS/Heavy-Rail-Track-Construction-Standards-Part-C.pdf

8.12 Summary

The purpose of this chapter was to:

- Understand the function of a track structure
- Define the terms used in railways
- Identify and discuss the various terms of a track structure
- Discuss the function of each element in a track structure
- Discuss railway line maintenance.

Self-evaluation 8

1. Complete the sentences:
   a. ______ refers to large aggregates underneath the sleepers providing support.
   b. Gauge is the ______ distance between two rails.
   c. Rail profiles used in South Africa are specified in ______ (units).
   d. Sleeper ______ will depend on the loads they carry but vary between 600 and 750 mm.
   e. ______ are used to hold the rails securely onto the sleepers and prevent undue movement.
2. State whether the following are true or false:
   a. Rails are made of concrete.
   b. Of the nine different rail profiles available in South Africa, five are not used today.
   c. A fish-plated joint makes use of a six-hole dry joint.
   d. Three types of railway sleepers are still used today.
   e. The formation refers to the foundation on which the track is constructed.
3. Answer the following short questions:
   a. What do you understand by the term 'flat' wheel (referring to railways)?
   b. Explain what you understand by a rail being specified as 48 kg/m.
   c. Why do you think grades on railway lines play such an important role?
   d. Describe the properties expected of rails.

Answers to self-evaluation 8

1. a. ballast
   b. horizontal
   c. kg/m
   d. spacing
   e. fastners
2. a. false
   b. false

- c. false
- d. true
- e. true

3. a. See ref 8.4, page 178
   b. See ref 8.4, page 178
   c. See ref 8.10, page 184
   d. See ref 8.4.1, page 178

Chapter 9

Airports

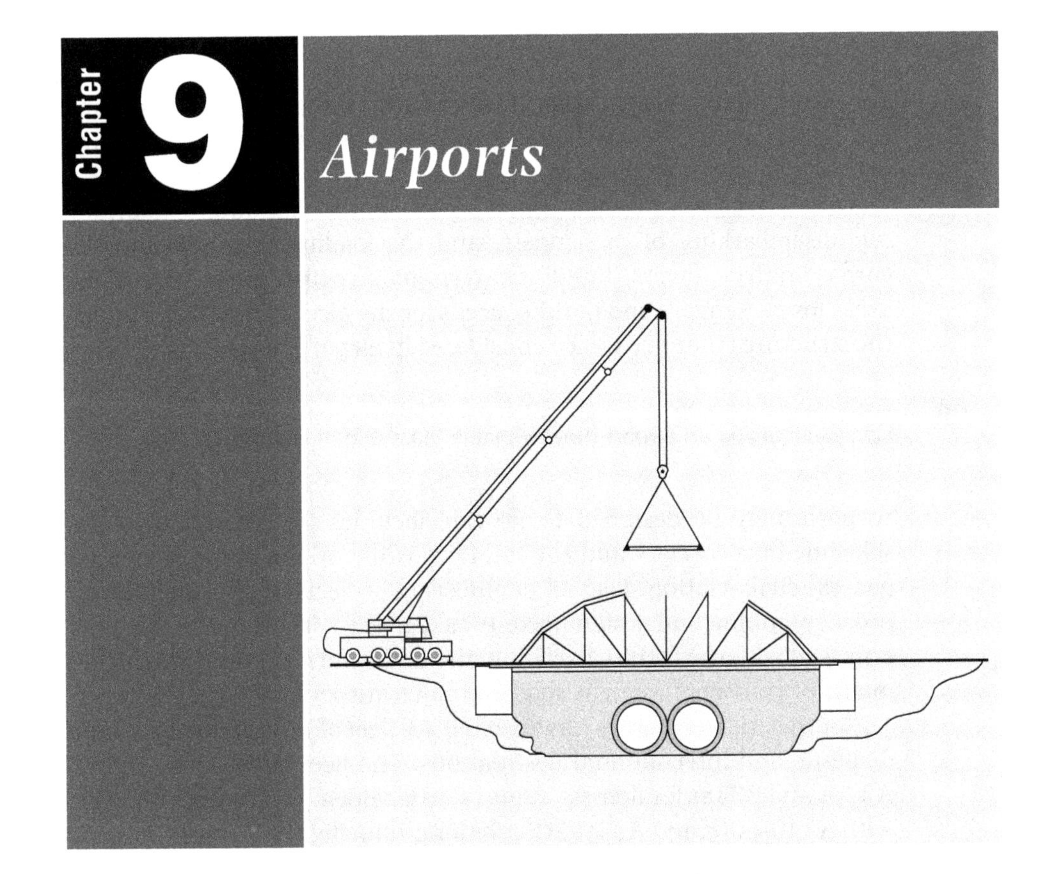

Outcomes

After studying this unit, you should be able to:

- Discuss the phases of airport design
- Identify and explain the elements of airport structure
- Discuss the importance of airport pavements
- Discuss the importance of drainage facilities.

9.1 Introduction

Air travel is probably the fastest way of travelling long distances, and the development of airports has grown to be an important part of the overall transportation system.

Large areas are needed to make provision for landing, embarking and disembarking of passengers, and the loading and offloading of cargo. Control towers, buildings (terminals) which serve as holding areas for passengers and parking areas for the aircraft, are just some of the structures that need to be considered in airport design.

Can you imagine an airport runway that is more than 4 km long?

Airports may be designed to provide facilities for general aviation, scheduled air carriers and the military, while others are designed for general civil aviation. Like all engineering projects of this magnitude, proper planning and design needs to take place. Airport design does not only deal with the facility itself, but also links and integrates with other modes of transport, such as road and rail transport.

South African Airways has an extensive fleet of aircraft flying to both domestic and international destinations. The fleet consists of A319s, A320s and B738s for domestic routes, while Boeing 747s were replaced with A340-600s and A330-200s for long-haul flights. To give you some indication of the statistics of these two aircraft i.e. one internal and one long-haul plane, look at the following table:

Table 9.1 Comparison of specifications for domestic and international aircraft

Specification	Airbus A320	A340-600
Overall length (m)	38	75.36
Wing span (m)	34	63.45
Height (m)	12	17.22
Empty weight (tons)	42	245
Loaded weight (tons)	77	369
Speed (km/h)	820	1 053
Passengers	150	380/419/475*

** Seating is given in three classes (first, business and economy). The first total is for an aircraft configured with all three classes, the second business and economy, and the last economy only.*

In everyday travel, flights accomodate 3 classes of travel and the occurrence of having a flight of only economy class travelers is extremely rare.

The A380 double-deck aircraft is the largest commercial aircraft flying, capable of carrying 525 passengers, but able to be if configured for economy, business and first class cabins, expanded to 853 if economy class only. The distance (range) that this aircraft can fly on a fully loaded fuel supply is approximately 15 700 kilometres.

The typical specification is outlined in the table 9.2.

Table 9.2 Typical specifications of the A380

Specification	Airbus A380
Overall length (m)	72.72
Wing span (m)	79.75
Height (m)	24.09
Empty weight (tons)	360
Loaded weight (tons)	560
Maximum fuel capacity	320 000 litres or 320 tons
Speed (km/h)	When cruising approximately 960km/h but can reach maximum speeds of up to 1100km/h
Passengers	525/853*

**The first total is with a three-class configuration and the second all economy.*

Have you ever wondered whilst watching an aircraft take off, what its speed was? It obviously will be influenced by various factors like:

- The size and engine capacity of the plane;
- The height above sea level of the airport (air pressure);
- Environmental conditions such as wind, snow, rain, etc.;
- Take-off weight of the aircraft; and
- Length of the runway.

The take-off speed of a large 747 aircraft can be anything between 260–370km/h, whilst landing speeds are approximately 280km/h.

Can you imagine an aircraft weighing 560 tons being kept in the air, let alone the forces transmitted into the ground when it lands?

9.2 Definitions

- A **terminal** is a building that receives passengers and cargo, and allows for their processing to take place.
- An **apron** is a hard-surfaced area in front of the terminal building where aircraft can stand or park.
- A **holding apron** is the area adjacent to the taxiway or near the runway entrance, where aircraft may park temporarily.
- **Taxiway** refers to the marked path along which an aircraft moves to or from a runway. It is similar to the runway but smaller in dimensions.

- **Airside** includes all the activities on the airport side of the terminal – for example, runways, aprons, holding areas, taxiways, etc.
- **Landside** refers to all the activities that are non-flying related – for example, access to and from the airport, parking, etc.
- **Runway** is a graded and paved load-bearing area allowing for aircraft take-off and landing. It is also commonly referred to as a landing strip.
- **Pier fingers** resemble fingers extending from the palm of a hand, allowing for increased airside activity. All processing is done at the main terminal; boarding takes place at pier fingers. An example of this airport layout is O'Hare International in Chicago, USA.
- **Pier satellites** are a move towards decentralisation of the pier finger concept, where they provide holding areas for passengers. An example of this airport layout is Stuttgart Airport, Germany.
- **Remote satellites** of a central terminal are connected by a mechanised form of transport. For example, in Los Angeles, USA, passengers are transported to the building from which the aircraft will depart.
- **Remote aprons** follow a centralised approach for processing and then transporting passengers to where the aircraft is parked on the apron. An example of this type of layout is at Dulles International in Washington DC, USA.
- **Remote piers** are a fairly recent innovation, where the central terminal is linked under the apron to remote piers. An example of this layout is Atlanta Airport, USA.

9.3 Airport master plan

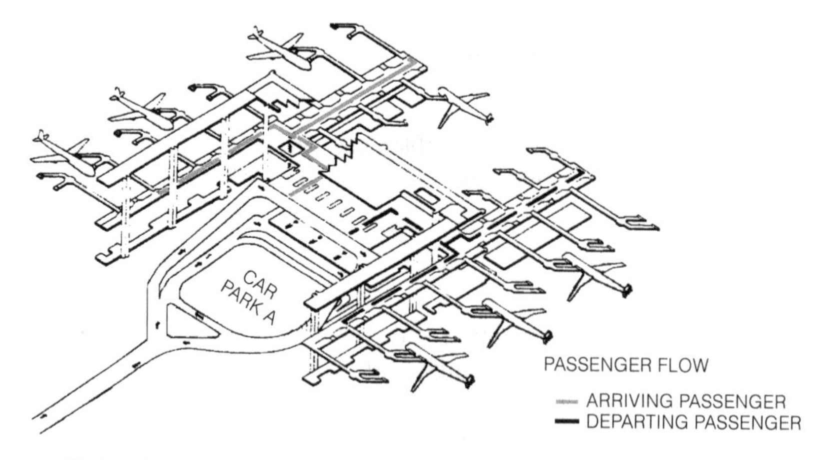

Fig 9.1 Changi Airport, Singapore, terminal building 1: passenger flow

Airports are designed according to a four-phase plan recommended by the Federal Aviation Authority (FAA).

Phase 1: Airport requirements

During the first phase, the forecasting of travel demands in relation to facility requirements is done to establish the magnitude or scope of the project. Once completed, but more often simultaneously, an initial environmental impact assessment is undertaken to determine whether there would be any detrimental effects arising from having the airport in that location.

Phase 2: Site selection

Once the size of the airport has been established, a suitable site is investigated. The site must be evaluated with respect to airspace requirements, access, a full environmental assessment, land cost and availability, site development costs and political implications.

Phase 3: Airport plans

After the site has been selected, careful planning of the actual facility takes place and includes the following:

- An **airport layout plan** showing all buildings, their sizes and location. This is similar to a layout plan for any other construction project.
- A **land use plan** that shows what the land that will form part of the airport and its surrounds will be used for.
- **Terminal area plans** are more detailed versions of the airport layout plan, showing the size and location of buildings within the terminal area.
- An **airport access plan** indicates links with other modes of transport.

Phase 4: Financial plan

This is a study of the economic feasibility of the airport, including its development and construction costs.

Once all four phases have been considered, the project will enter the implementation phase.

9.4 Airport structure

Airports consist of several facilities, some of which will be discussed below.

9.4.1 Runways and taxiways

A **runway** is the paved load-bearing area that aircraft use to land and take off. Its size (length and width) is dependent on a variety of factors, but can range between two and four kilometres in length and 20 to 100 m wide.

Did you know that the number of aircraft expected during a given period determines the number of runways at an airport?

Because of the technological advances in aircraft design and construction resulting in planes becoming larger and able to carry additional loads, these factors must be taken into account when designing the length, width, clearances, grades, etc. of a runway.

Several runway configurations exist, but most runway systems are arranged according to a combination of four basic configurations:

- Single runways;
- Parallel runways;
- Open-V runways; and
- Intersecting runways.

Fig 9.2 Intersecting runways, Heathrow Airport, London, UK

Fig 9.3 Parallel runways, Changi Airport, Singapore

Do you think the wind plays an important role in the location of the runway?

You may have noticed that planes land and take off into the wind and that the predominant wind direction determines the direction of the runway. Crosswinds are also taken into account when designing the width of the runway.

The capacity of a runway or runway system depends on a number of factors, including:

- aircraft types – large aircraft require long, wide landing areas;
- performance characteristics of the aircraft – for example, the length of runway required for take-off and landing;
- landing and take-off gross weight of aircraft – the heavier an aircraft, the longer it takes to become airborne;

- air traffic control techniques – the control of aircraft on and off the ground;
- apron capacity – a small apron will only be able to accommodate a limited number of aircraft;
- landing aids – for example, navigational aids, fire-fighting equipment, runway lighting, etc.;
- landing system in use (visual or instrumental landing systems);
- elevation of the airport; and
- runway gradient.

At busy airports there will probably be an additional runway parallel to the main runway, which is used as a taxiway for aircraft that are either ready to take off or have just landed. The taxiway is used to connect the runways to the terminal apron and should always lead onto the end of the runway. This aligns the aircraft with the centreline of the runway, in preparation for take-off. At very busy airports, you may find more than one taxiway to cater for the increased activity.

The volume of air traffic, the runway configuration, and the location of terminal buildings and other ground facilities determine the number of taxiways at an airport.

Activity 9.1

Why do you think the elevation of the airport has an influence on the runway system? Does thinner air at higher altitudes have anything to do with it? What influence does air temperature have on runways? Explore these questions in the library or by using the internet and discuss your findings in class.

9.4.2 Airport aprons

An **airport apron** is a defined area intended to accommodate aircraft for the purpose of loading and unloading passengers or cargo, refueling and maintenance.

There are two types of apron referred to in airport design: a **terminal apron** and a **holding apron**. The terminal apron is usually located near the terminal (often in front of the terminal building), but at larger airports they may be located away from the terminal area.

Holding aprons are areas where aircraft can park briefly while flightdeck checks and engine run-ups are made before take-off clearance is given. Holding aprons are required where aircraft operations exceed 30 per peak hour.

The planning and design of the aprons depend on:

- the terminal configuration;
- the movement characteristics of the aircraft at the airport;
- the physical characteristics of the aircraft (size, wingspan, gross weight, etc.); and
- the type and size of ground service equipment (e.g. fire-fighting equipment).

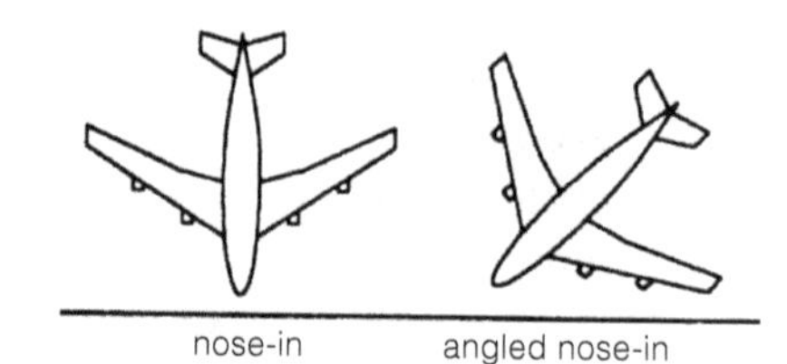

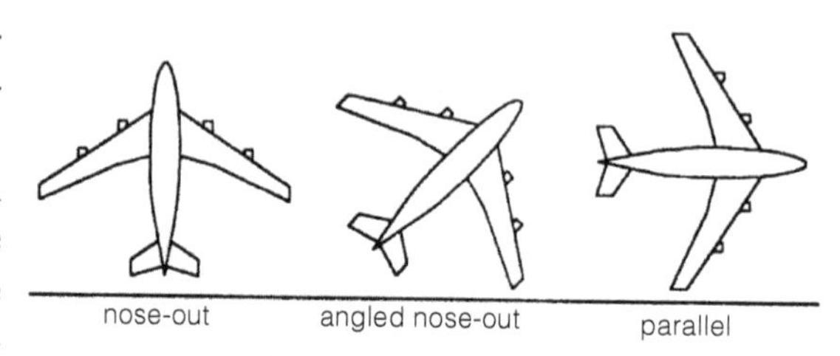

Fig 9.4 Aircraft parking configurations

The form of airside interface (activities relating to the number and movement of aircraft) and the design dimensions of the apron depend on the number of gates and the parking configuration chosen. The five parking configurations used at airports are: nose-in, angled nose-in, nose-out, angled nose-out and parallel.

9.4.3 Terminal buildings

The design and location of terminal buildings can vary and include several options:

- The building can be located between two parallel runways;
- It can be located in the angle of two convergent but non-intersecting runways; or
- It can be in the centre of a complex of parallel runways.

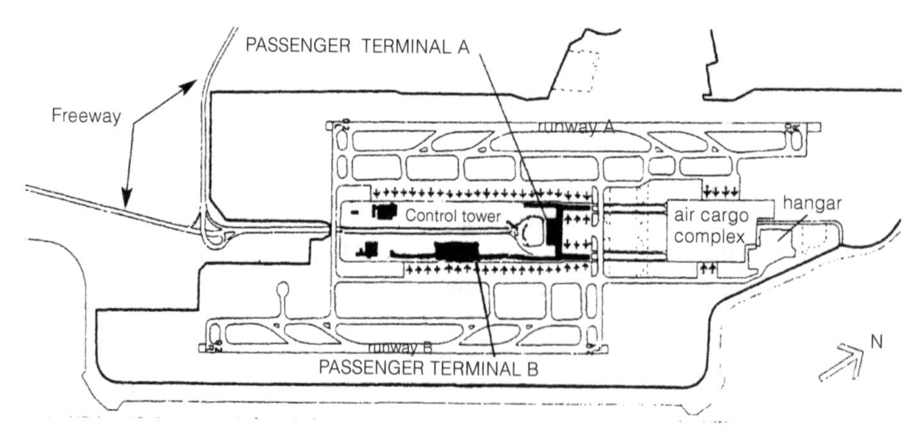

Fig 9.5 Typical airport layout

The functions of the terminal buildings include:

- **processing** passengers and cargo (tickets, baggage, etc.);
- a **change of transport mode** (to car, taxi, bus, train, etc.);
- **holding areas** for **transfers** (departure lounges, car rental, service areas, etc.); and
- **airline** and **support activities** (airline offices, management, public address system, etc.).

Entertainment and shopping facilities are also provided at most large airports around the world.

Arrival and departure areas are usually found at different locations to speed up passenger processing. For efficient and convenient passenger movement, the buildings should be well grouped to facilitate transfers from one passenger line to another.

Terminal designs have developed to such an extent that some airports have moved away from a centralised processing system and adopted a more decentralised approach. Decentralisation involves spreading the passenger processing functions over a number of centres in the terminal complex. Terminal configurations to support this approach are pier satellites, pier fingers, remote satellites, remote aprons and remote piers.

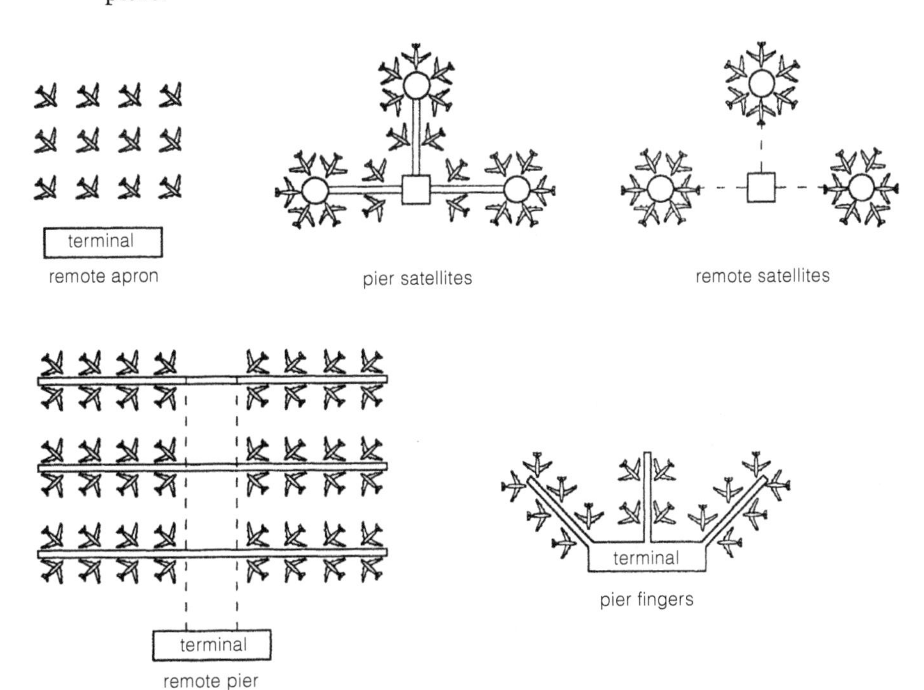

Fig 9.6 Terminal configurations

9.4.4 Hangar and service buildings

Hangars that can accommodate the largest jets have to be built for undertaking servicing and repair work. Buildings are also provided for everyday requirements of an aircraft, such as food, beverages, cleaning and minor service facilities.

9.4.5 Airfreight buildings

Airfreight is becoming a more viable means of transporting goods, especially those that are required urgently. The airport must have a building that can process and store incoming and outgoing freight.

9.4.6 Parking

The most popular means of travel to and from an airport is by car, bus or taxi. Some of the more modern airports have a rail or underground rail network that supports passenger movements. The Gautrain to OR Tambo Airport and the underground rail to Heathrow Airport in London are examples of such facilities.

Besides the external movement of passengers (i.e. passengers coming to and from the airport), there are also some airports who use rail transport to move passengers between terminals. Provision must be made to accommodate the influx of these other modes of transport. At many airports, public and private transport facilities are located in different areas of the airport. Distinction is also made between short and long-term parking and drop-off zones.

The area needed to accommodate cars is extensive, when one considers that there are six main parking categories:

- Kerbside parking for setting down and picking up passengers;
- short-term parking (up to 48 hours);
- long-term parking;
- staff parking (for airport and airline staff);
- spectator parking; and
- car rentals.

The exact location of these parking areas will vary according to the layout of the airport and the terminal buildings.

9.5 Pavement structure

To withstand the enormous loads applied to the ground, it is necessary to provide suitable pavements for aprons and maintenance areas, and for runways and taxiways.

Design criteria include:

- adequate strength for all current and future aircraft likely to use the airport;
- adequate fatigue strength (to withstand repetitive loading of aircraft movement and landing);
- absence of loose particles that could be sucked into jet engines;
- resistance to jet blast (the high-velocity, concentrated stream of air resulting from aircraft engines);
- resistance to fuel spillage (certain fuels, like diesel, tend to soften especially bituminous surfacing);
- ability to withstand temperature fluctuations;
- good surface drainage;
- good skid resistance;
- good riding surface; and
- ease of maintenance.

Generally, three types of pavement structures are used at airports: rigid, flexible and composite.

9.5.1 Rigid pavements

Concrete (rigid) pavements act like foundations, spreading the load over a wide area of the subgrade. The concrete is usually unreinforced and divided into rectangular panels or bays to restrict the tensile stresses to which it is subjected.

Contraction joints separate bays and the bay size depends on the slab thickness. To improve the skid resistance of the concrete surface, the concrete may be wire combed, or small transverse grooves may be made in the surface while the concrete is still plastic. Alternatively, the hardened concrete may be scored with diamond cutting drums.

Concrete pavements are generally used for aprons where sustained loads over longer periods (due to parked aircraft) can better be supported by rigid pavements.

9.5.2 Flexible pavements

A flexible pavement at an airport behaves in a similar manner to that of a road. It relies on its thickness and elasticity to disperse the high loads and minimise the stresses on the subgrade material. A flexible pavement is made up of a number of layers of granular material that increases its rigidity and decreases its flexibility towards the surface. The riding surface or wearing course should be an impervious, asphalt layer.

Well-designed flexible pavements have good riding qualities, but some surfaces are susceptible to jet heat and fuel spillage that can cause surface softening.

The high landing and take-off speeds of modern aircraft, combined with flat transverse slopes on runways, have led to the problem of aquaplaning. When water falls onto a flat surface, it tends to form a film and can result in tyres losing contact (traction) with the surface and skidding or aquaplaning. Friction coatings on flexible runways, and grooving, wire combing or scoring on rigid pavements are used to alleviate this.

What other environmental conditions are airplanes faced with when taking off or landing? How do you think this impacts both the plane as well as the design of the airport facility, runways, etc.?

9.5.3 Composite pavements

In a continuous concrete pavement, cracking accentuated by exposure to heavy traffic is likely to develop, regardless of the quantity of reinforcing that has been incorporated. However, if a bituminous surfacing overlies the concrete pavement, cracking is reduced since the variation in the concrete's temperature is lowered. Cracks that form in the concrete but are not subject to wear are unlikely to be severe.

9.6 Drainage

As with any engineering facility, it is very important to consider and cater for the water, whether surface or subsurface, on site. We have already looked at the damaging effects of water and it is therefore necessary to minimise this by providing drainage.

9.6.1 Surface water drainage

To protect the subgrade from the ingress of water, it is essential to have an effective surface drainage system. We already know that, when water penetrates the subgrade, it becomes soft and loses its strength. Poorly drained subgrades have a greatly reduced bearing capacity and may cause failures in pavement areas. Poor surface drainage will decrease the skid resistance and cause aquaplaning.

9.6.2 Drainage from runways, taxiways and aprons

Gullies, continuous gratings and slot drains are used to drain surfaced (paved) areas into a proper drainage system. Ditches are not allowed adjacent to areas containing air traffic as they are safety hazards, but may be used to carry away surface water remote from aircraft operations.

9.6.3 Drainage from margins

The main requirement of margin drainage is to prevent water ponding near areas of aircraft movement. To achieve this, surface areas are gently graded and subsoil drains are provided. Cut-off drains are necessary where the ground slopes towards pavement areas.

9.6.4 Subsoil drains

To prevent the water table from affecting the subgrade, subsoil drains are provided. These may also be used for draining low-lying, waterlogged areas. There are three subsoil drain installation patterns: herringbone, parallel and gridiron. Open-jointed porous pipes are laid to depths ranging from 600 mm to 1 200 mm and connected to the drainage system.

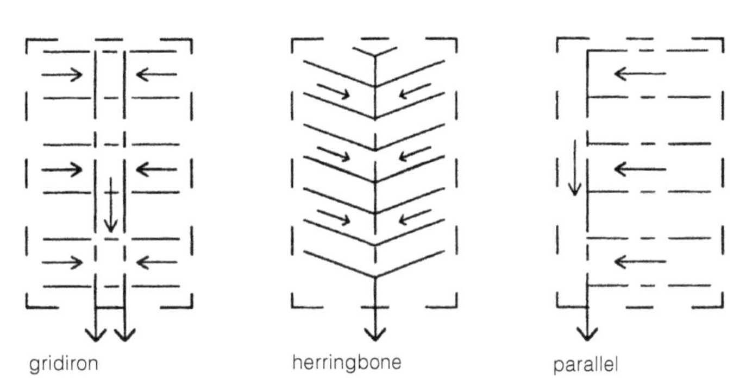

Fig 9.7 Subsoil drains

Activity 9.2

Visit an airport and see if you can identify the structures described in this unit. Draw sketches or take photographs to support your findings. Make a list of items to justify why you think the airport is located in this area or motivate why it should be located elsewhere. Consider the following aspects:

- availability of sufficient land area and land use;
- accessibility;
- links with other transport facilities;
- topography;
- obstructions (natural and man-made); and
- benefits to local communities.

Debate your findings in class.

Why do you think Durban International Airport was re-located to its current location? Why not 'Google' the answer and discuss your findings with your classmates?

If you are interested in finding out more about aviation, look at the following websites:
- http://www.saa.co.za
- http://www.airbus.com
- http://www.letsfindout.com/subjects/aviation

9.7 Summary

The purpose of this chapter was to:
- Discuss the various phases in airport design
- Identify and explain the various elements of airport structures
- Discuss the importance of airport pavements
- Discuss the importance of drainage facilities.

Self-evaluation 9

1. Complete the sentences:
   a. _______ refer to the paved load-bearing areas used for aircraft landings and take-offs.
   b. An _______ is the defined area used to accommodate the loading and unloading of passengers and cargo.
   c. A _______ _______ is an area where aircraft park temporarily while awaiting take-off clearance.
   d. _______ refers to a thin layer of water on a surface that can result in loss of tyre traction.
2. State whether the following are true or false:
   a. An airport layout plan is a detailed plan showing the size and location of buildings within the terminal area.
   b. Heavier aircraft take longer to become airborne.
   c. There are two types of aprons in airport design.
   d. Composite pavements comprise rigid and flexible pavement structures.
3. Answer the following short questions:
   a. What is the difference between a taxiway and a runway at an airport?
   b. What is an airport apron used for?
   c. List the functions of airport terminal buildings.
   d. What is a composite pavement structure at an airport?

Answers to self-evaluation 9

1. a. runways
   b. apron
   c. holding apron
   d aquaplaning
2. a. false
   b. true
   c. true
   d. true
3. a. See ref 9.2, page 189
   b. See ref 9.4.2, page 193
   c. See ref 9.4.3, page 194
   d. See ref 9.5.3, page 198

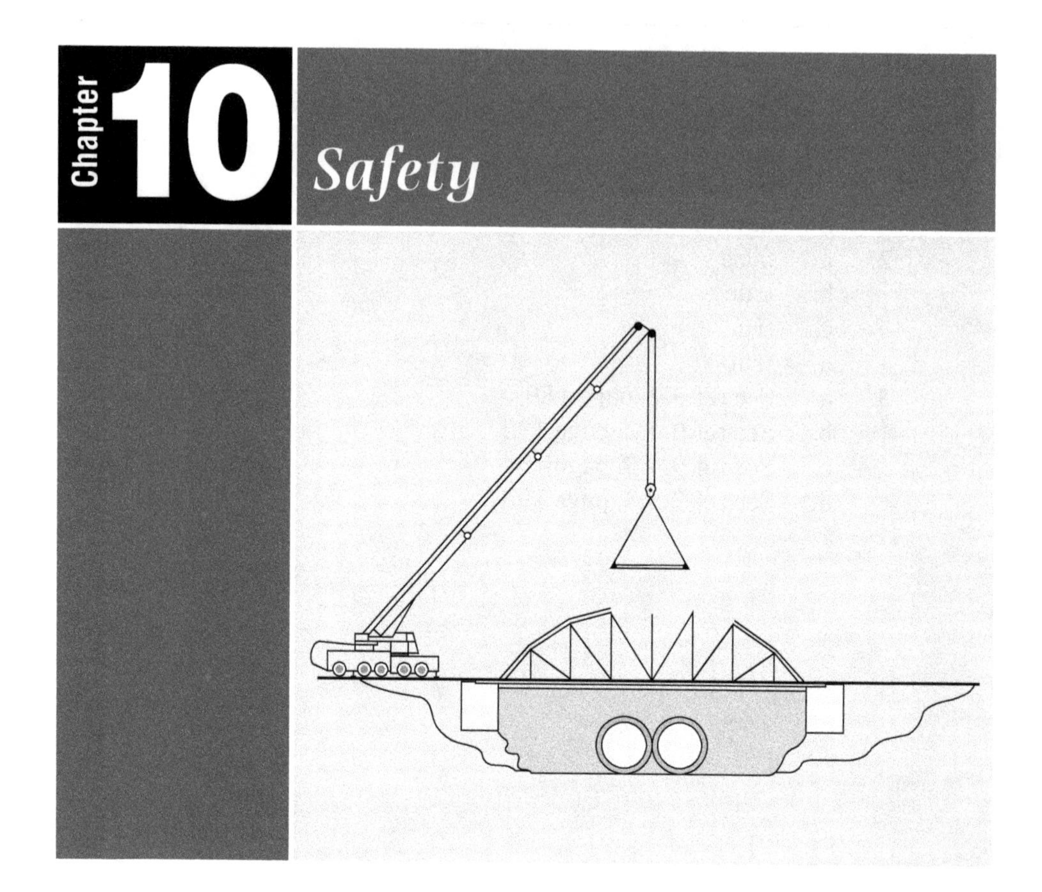

Chapter 10 Safety

Outcomes

After studying this unit, you should be able to:

- Discuss the role of safety in construction
- Identify various procedures to ensure a safe working environment
- Identify the various Acts relating to safety and the workplace.

10.1 Introduction

The construction industry is regarded as one of the most hazardous industries in the world. It has been statistically proven that, on average, construction workers sustain two times more injuries and three times more fatalities than other workers. One of the most important measures of safety performance is workers' compensation insurance.

The poor safety record in the construction industry is not only limited to human suffering. It also contributes to other costs, such as productivity, company reputations and rescheduling delays.

Safety in the work environment is particularly important considering that on average a worker spends approximately nine hours at work per day.

The South African government, heeding the call for safer work environments, has implemented several Acts in an attempt to reduce high accident rates. Government makes the laws to which management must adhere. Management must ensure that the work environment is safe from any undue hazards, and workers must adhere to the rules and regulations provided. In many instances, accidents and injury result from unsafe acts by workers.

We all have a role to play in ensuring a safer work environment.

Certain laws have been passed by parliament to ensure workers work under conditions that protect their well-being.

10.2 Machinery and Occupational Safety Act

The primary objective of this Act is to control the working conditions of all workers in industry and commerce, with the exclusion of those in the mining industry. The Act helps to prevent accidents and industrial diseases. To achieve the objects of the Act, an employer must ensure that:

- machines used in the workshop are fitted with safeguards;
- protective equipment and clothing are provided for workers;
- workers are properly instructed on how to operate machines safely; and
- the workplace is free from steam, smoke, noxious fumes and gases that can cause illness and disease.

Employees also have certain responsibilities for safety under the Act. They must:

- use all the safety equipment supplied;
- not interfere with or abuse safety equipment;
- do nothing willful to endanger the health and safety of themselves or others; and

- obey orders issued in the interests of safety either by a superior or in the form of a general standing order.

To ensure safety instructions are followed by the workers at all times, the Act requires that the employer delegate certain safety functions to competent staff members.

The Act requires that employers appoint safety representatives and establish safety committees subject to certain conditions. The safety representative must be a full-time employee of the company and should have a reasonable standard of competence and knowledge. He or she must be able to submit full reports to safety committees and inspectors.

The Act states that a person working on premises where machinery is used, shall immediately report to his or her superior anything which comes to his or her notice that is liable to cause danger to persons or accidents involving machinery.

No person shall consume, or offer to any other person, or have in his or her possession intoxicating liquor while in the vicinity of or while working on or near machinery. No person in a state of intoxication shall enter or remain on the premises where machinery is used.

The Act requires that all employers keep an accident register recording:

- the date and time of accidents;
- the name of the injured person;
- a description of the accident;
- when the divisional inspector of labour was notified of the accident; and
- when work resumed.

An inspector must be notified when an accident occurs in a factory or on premises where machinery is used, or where excavation is performed. The accident must be one that results in:

- the death of a person;
- an injury that is likely to cause death;
- the person losing a limb, part of a limb or sustaining a permanent physical defect; or
- a person being unable to continue his or her normal activities for 14 days as a result of an injury.

An inspector must also be notified if a person is rendered unconscious as a result of heat stroke, heat exhaustion, electric shock or the inhalation of fumes or poisonous gas.

Management must provide an adequately stocked first-aid box if it employs five or more people. A qualified first-aider must be employed for staffs of 10 or more. A separate, properly equipped first-aid room must be provided where 100 or more people are employed.

Any person who fails to comply with requirements of the Act will be guilty of an offence and, if convicted, could be liable for a fine not exceeding R2 000, or imprisonment exceeding 12 months, or both.

10.2.1 Responsibilities of a builder in supervising building work

Every structure to be built shall be under the supervision of a responsible person, who shall be a competent person and who shall be appointed by the builder in writing. This responsible person shall be in charge of all the building work and ensure that:

- the provisions of the Act are complied with;
- all plant and machinery is maintained in good condition and properly used; and that
- all work is carried out in a safe manner and in accordance with the designs and specification as approved by the appropriate authority.

10.2.2 General safety measures the builder must carry out

All stairways, passageways, gangways, basements and other places where danger may exist through lack of natural light, must be adequately lighted.

All stairways, passageways and gangways, where practicable, must be kept free from materials, waste or any other obstructions.

All openings in floors, hatchways and stairways, or any other opening from which persons are liable to fall, must be adequately boarded over, fenced or enclosed with suitable rails or guards to a height of not less than 900 mm and not more than 1 100 mm from the ground or floor, provided that such boarding or guarding may be omitted or removed for the time and to the extent necessary for the access of persons or the movement of materials.

A suitable catch platform or net must be erected above a place where persons regularly work or pass, or the danger area must be adequately fenced off, especially if work is being performed above such entrance, passageway or place and there is danger of persons being struck by falling material.

A scaffold plank of timber must be at least 228 mm wide by 38 mm thick; it must rest on at least three supports and not project more than 230 mm at the end supports. Boards must be fastened to the scaffolding to prevent displacement.

The builder must provide for suitable enclosures on all sides of the tower at ground level, where persons are liable to be struck by moving parts of a hoist, except on the side that gives access to the platform.

Gates or doors must be provided at each floor landing and must be kept closed, except when the hoist platform is at rest at that landing.

Excavation that is accessible to the public must be fenced off with a barrier of at least 600 mm high and be provided with a red light at night. When excavations are deeper than 1.5 m, builders must assure and grade the sides properly before a person can work in the excavations. This, however, is not required where the excavation is in solid rock or where the sides of the excavations are sloped to the natural angle of the earth.

A notice in the official languages of the particular province prohibiting unauthorised persons from entering the premises, must be posted up in a prominent place at the entrance to the building site.

When a worker is exposed to danger, the builder or excavator must provide, free of charge, and maintain in good order, adequate protective clothing, appliances and material.

Dangerous goods, such as inflammable liquids and explosives, must be stored in strict observance to regulations.

10.3 Protective clothing

General physical protective equipment available:

- **Eyes:** goggles, spectacles, face masks and face shields
- **Head:** protective hats (hard hats), caps and hair nets
- **Body:** aprons, suits and jackets, kidney belts and safety belts
- **Legs:** spats (covers insteps and ankles) and leggings
- **Hands:** gloves and mittens
- **Feet:** safety boots and shoes
- **Ears:** earmuffs and earplugs
- **Lungs:** respirators and face masks

One of the basic problems of safety is getting workers to use the equipment. Factors that influence the worker in using the equipment include:

- Can the equipment be worn with ease and comfort?
- Does it interfere with normal working procedures and operations?
- Does the worker understand why it is necessary to use the equipment?
- Can economic or social considerations or disciplinary action be used to make the worker use protective equipment?

10.4 Workman's Compensation Amendment Act

The main functions of the Act are:

- paying compensation to workers for industrial accidents and occupational diseases, and meeting the medical costs involved;

- preventing industrial accidents; and
- rehabilitating injured workers.

When a worker is injured, he or she is entitled to compensation for the period he or she is laid off. Should the worker be killed, dependants are entitled to compensation. No compensation is paid if the worker is guilty of serious and willful misconduct, unless the accident causes serious disability or the worker dies. If the accident leads to serious disability or death, the Commissioner will pay compensation, even if the worker contravened the law or the employer's instruction, or acted without instructions, but only if he or she was injured while engaged in the employer's business.

Serious and willful misconduct means:

- drug use;
- drunkenness;
- contravening any law or regulation made to ensure the safety or health of workers, or with reckless disregard; and
- any other act or omission which the Commissioner considers to be serious and willful misconduct.

A worker is not entitled to compensation until he or she sets foot on the workplace premises. Thus, workers are not covered while travelling between home and work, although there are exceptions.

If a person is called out at night for work purposes, he or she is covered while travelling from home to the job and back again. Other types of workers, like insurance agents, who usually operate from home, are likewise covered once they leave home. If company transport carries staff to work, then all workers are covered once they leave home. If company transport carries staff to work, workers are covered once they enter the vehicle.

The Commissioner can lower the assessment on employers who design, equip, organise or conduct their business in such a way as to prevent accidents. Here the drain on the accident fund should be lower than the average for other firms in that class.

The Commissioner, at his discretion, may calculate the employer's assessment on a lower percentage of his annual wage than that used for other employers in the industry. If, however, the reverse position were to occur, the Commissioner may assess the employer on a higher percentage of his annual wage.

10.5 National Occupational Safety Association (NOSA)

NOSA is a company registered under the Companies' Act as an association not for gain. It was established in 1951 by a major employer organisation in conjunction with the Workman's Compensation Commissioner. The association's aims are to supply a service to management in industries by advising on accident prevention techniques and to promote safety by means of talks, pamphlets, posters, literature and safety training courses.

Employers are reassessed annually to work out what their contribution to the accident fund should be. The contribution is based on the number and severity of accidents. The accident fund is administered by the Workman's Compensation Commissioner under the Workman's Compensation Act. The Commissioner may grant money to NOSA to carry out its basic functions. There are essentially four basic causes of accidents:

- unsafe conditions;
- lack of knowledge and/or skills;
- physical or mental defects; and
- improper attitude.

A commonly used word in the building and civil engineering industry is housekeeping. This means that there should be a proper place for everything at all times. It is important because:

- it reduces the time spent on finding and checking goods, articles and tools;
- it saves space when everything is packed away tidily;
- injuries are avoided when gangways and working areas are kept clear of unnecessary materials; and
- fire hazards are reduced if combustibles are kept in proper containers.

If a contractor does not practice good housekeeping, he or she can expect the following types of accidents:

- workers tripping over loose objects;
- articles dropping from above;
- workers slipping on greasy, wet or dirty floors;
- staff running against projecting, badly placed materials;
- hands or other parts of the body cut on projecting nails, steel strapping, splinters, wire, etc.; and
- fires.

Good housekeeping will ensure a lower risk of accidents and improve the organisational aspects of any company.

What is housekeeping?

10.6 Summary

The purpose of this chapter was to:

- Review and identify the different Acts applicable to safety in the workplace
- Explain the importance of a safe working environment
- Discuss the processes involved in ensuring a safer workplace.

Self-evaluation 10

1. Complete the sentences:
   a. Machines used in workshops must be fitted with _______.
   b. _______ must provide an adequately stocked first-aid box if it employs five or more people.
   c. A scaffold plank must rest on at least _______ supports.
   d. _______ in industrial terms means that there should be a proper place for everything at all times.
   e. Fire hazards are reduced if _______ are kept in proper containers.
   f. NOSA supplies a service to management by advising on ___________ techniques.
2. State whether the following are true or false:
   a. Workers have a responsibility to ensure a safe working environment.
   b. Government has no role to play in reducing accidents at work.
   c. Safety functions must be delegated to competent staff.
   d. It is not necessary to provide lighting near dark stairways.
   e. One of the main functions of the Workman's Compensation Act is to prevent industrial accidents.
   f. When a worker is intoxicated, he or she is entitled to compensation after an accident.
3. Answer the following short questions:
   a. What are the responsibilities of employers with regard to safety?
   b. What information must be recorded after an accident has taken place?
   c. What is considered to be serious and willful misconduct?
   d. What are the basic causes of accidents?

Answers to self-evaluation 10

1. a. safeguards
   b. management
   c. three
   d. housekeeping
   e. combustibles
   f. accident prevention
2. a. true
   b. false
   c. true
   d. false
   e. true
   f. false
3. a. Read the unit
   b. see ref 10.2, page 203
   c. see ref 10.4, page 206
   d. see ref 10.5, page 208

Chapter 11

Drainage

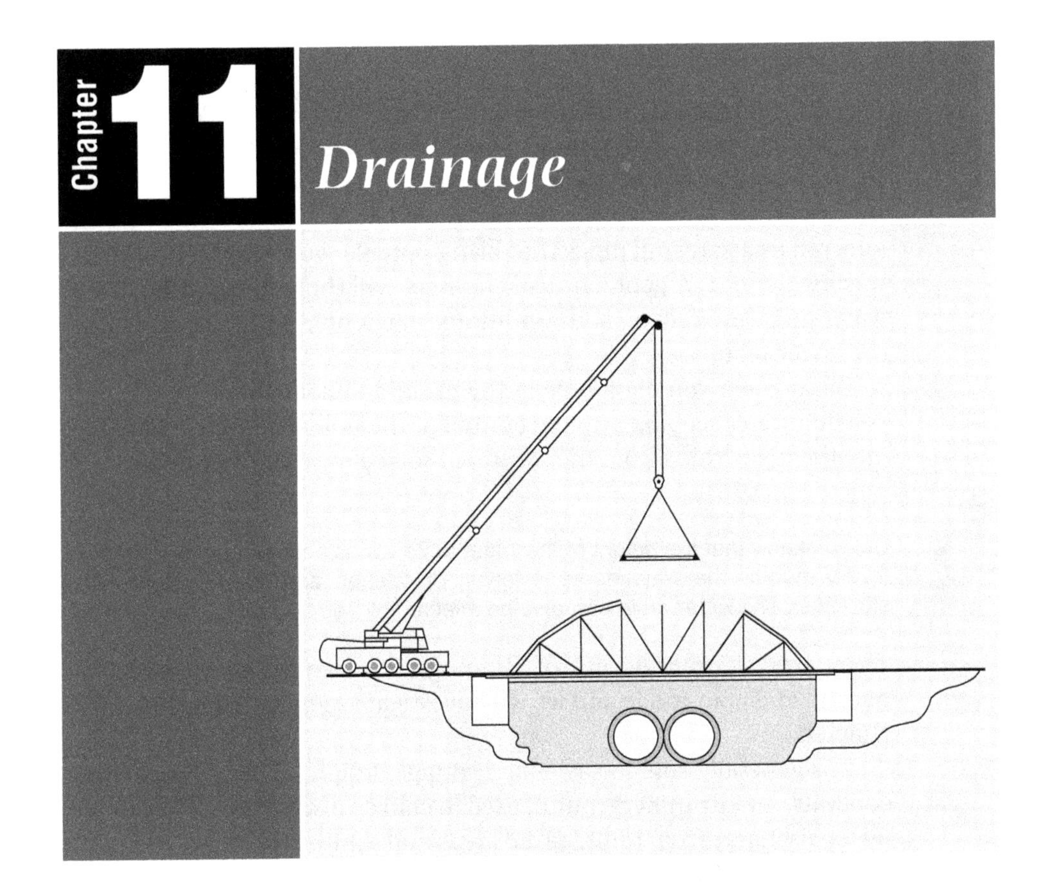

Outcomes

After studying this unit, you should be able to:

- Demonstrate the importance and application of various drainage systems
- Identify drainage system components and the materials from which they are made
- Explain testing methods used on pipelines
- Discuss the design process of a typical drainage system
- Discuss the importance of applying good drainage practice.

11.1 Introduction

The term 'drainage' has a variety of interpretations. For example, a civil engineer will think about drainage in terms of roads (surface water) or underground pipes (water mains, foul sewer and stormwater). A housing contractor will relate to drainage as the internal and external pipes that bring potable water into a house and wastewater out. A geotechnical engineer will think about sub-surface water and the effect on soil conditions, particularly the rise and fall of the water table.

Often when you dig a hole in the ground and you dig deep enough, you find water seeping into the bottom of the hole. The water that seeps into the hole drains from the saturated soil below the water table.

?

Did you know that the height of the water table varies according to the season? In dry months the water table is lower, or deeper, and in wet months it is shallower. The water table can also be affected by sea tides.

There are different forms of drainage: surface water, sub-surface, natural and man-made and we will investigate some of the forms in this chapter.

It is generally safe to say that drainage design involves a system of conduits, mainly underground, used to convey discharge – for example, water discharge from roofs, paved areas and sanitary fittings – to a point of discharge or treatment.

These pipes are available in a variety of sizes, shapes and materials. For example, there are concrete pipes, clay pipes and polyvinyl chloride (PVC) pipes, which are generally used in housing construction and drainage. There are also high density polyethylene (HDPE) pipes that are now more commonly used in water pipelines. Subsurface drainage pipes used to control ground water are also made of HDPE wrapped with a geotextile to provide a combination of filtration and strength.

Drainage pipes range from 40 mm diameter to as much as 2 m in diameter, which is large enough for a person to walk through. These pipes usually convey water from large reservoirs to urban areas.

In South Africa, different pipes are used for different situations. Foul sewer, also commonly referred to as sewage (wastewater and toilet water), is conveyed through PVC pipes or vitro clay pipes. A network of sewer pipes is called a sewerage system. Stormwater pipes are usually made of concrete and take rainwater from channels at the side of the road. Water pipes convey drinking water, commonly referred to as potable water.

Pipes transport water for domestic use and for use in manufacturing processes, and discard water by taking it to a place where it can be safely disposed of. A civil engineer designs pipe systems.

Some of these designs involve pipes that transport water by means of gravity (from a high point to a low point) and others involve pipes through which water is pumped from low points to higher points. To achieve this, pressurised mains are incorporated into the design.

In gravity fed mains, you will find manholes positioned either on the sidewalks or in the road. You may also find manholes in backyards. These manholes serve several purposes. They can be used for maintenance access by municipal workers in the event of a blockage in or damage to the pipe. Manholes can also be used by engineers or maintenance people to check that the lines are clean and intact. Foul sewer manholes are especially dangerous to work in as harmful gases are emitted by the sewage that flows in the pipes. Inhaling these gases can result in death.

? Have you noticed that municipal workers tie themselves to a harness before entering a manhole? If anything goes wrong, it is easy for them to be pulled to safety.

The arrangement of any drainage scheme is governed by:

- the internal layout of connections;
- external pipe positions;
- the relationship of one building to another;
- the location of public pipes (sewer and stormwater) and water connections; and
- the topography of the area to be served.

Gravity drainage systems must be designed within the limits of the drain so that the discharge can flow by gravity from the point of origin to the point of discharge.

The pipe sizes and gradients must be selected to provide sufficient capacity for maximum flows and adequate self-cleaning velocities at minimum flows to prevent deposits. In other words, the pipes must not be designed where the gradients are too small or the flow inside the pipes is too little, which will result in deposits of solid matter and blockages. Remember the slower water flows, the more likely it is that suspended matter within the water will settle to the bottom, resulting in deposits at the bottom of the pipe or channel. This may become problematic as the more settlement you have, the smaller the diameter and therefore the capacity of the pipe becomes – in other words, the pipe cannot carry as much water as it was originally designed for.

It is expensive to excavate and the deeper the excavation, the more expensive it becomes. It is therefore essential to consider the depth at which pipes will be placed. Other factors can also play a role. For example, the more rock there is, the more difficult it will be to excavate.

Similarly, the higher the water table, the greater the possibility that seepage into the pipes will occur. It is important to include infiltration of ground water into your design as it results in additional flows in your pipeline. If the terrain where the pipes are to be laid is hilly, it will be easy to design pipes that are gravity fed. In flat areas, drainage gradients could be problematic and a pumping system may need to be installed.

11.2 Definitions

- **Gravity** is the force of attraction that moves or tends to move bodies towards the centre of the earth.
- The **water table** is the surface (top) of the water-saturated part of the ground.
- **Conduits** are pipes or channels used to carry fluids.
- A **reservoir** is a natural or artificial lake for collecting and storing water for domestic or community consumption.
- **Dams** are barriers of concrete or earth built across a river to create a body of water.
- **Potable water** is water fit for human consumption (drinkable water).
- **Sewage** refers to waste matter from domestic or industrial areas that is carried away in sewers or drains.
- **Sewers** are underground drains or pipes used to carry waste matter (sewage) from domestic or industrial areas.
- **Sewerage** is a system or arrangement of sewers.
- **Jointing** is a junction of two or more pipes.
- **Invert** level is the lower, inner surface of a drain or sewer.
- **Manholes** are shafts with removable covers that lead down to sewers or drains through which a person can enter. Manholes are often referred to as inspection chambers.
- **Bore** refers to the diameter of a circular hole or pipe.
- **Hydrological study** is the study of the distribution, conservation and use of water.
- **Hydraulic study** is concerned with liquids in motion and the pressures exerted in this motion.
- **Nomogram** is a graph consisting of three lines that are graded for different variables so that a straight line intersecting all three gives the related values of these variables.
- **Catchment area** refers to the overall area (measured in square metres or kilometers) over which the influence of drainage will be experienced (for example, the area that drains to a river).
- **Storm duration** refers to the length of time rain falls.
- **Storm intensity** measures the amount of water that falls when it rains.

11.3 Drainage materials

Drainpipes are either rigid or flexible according to the material used in their manufacture. Clay is often used for rigid drainpipes in domestic work. Cast iron is an alternative. The usual materials for flexible drainpipes are pitch fibre and unplasterised PVC.

11.3.1 Vitrified clay pipes

Vitrified pipes are, as the name suggests, made of clay and 'baked' in ovens. Vitrified clay (VC) pipes are very popular in drainage. These pipes are no longer used, mainly due to their brittleness (they tend to break more easily than, for example, plastic pipes) and are replaced with other types of pipe wherever the need arises.

Specifications

There are two standard specifications for clay pipes:

- **SANS 559: Vitrified clay sewer pipe and fittings.** This specification covers all clay pipes and fittings and their physical properties, including dimensional tolerance, soundness, ovality, etc. It also sets the requirements for crushing strength and water absorption, sampling and testing. The minimum crushing strengths are (according to pipe classification):
  Class I 22 kN/m^2
  Class II 45 kN/m^2

- **SANS 974: Rubber joint rings (non-cellular).** This document defines the type of rubber and its specification to be used in the manufacture of joint rings – for example, hardness, tensile strength, acid resistance and joint strength.

Code of practice

- **SANS 058: The installation of sewerage and drainage non-pressure pipe lines.** This is a document used by the South African Bureau of Standards (SABS) which produces the South African National Standards (SANS) which serve as a guideline when installing and inspecting pipelines.
- **SANS 1200 series: Standardised specification for civil engineering construction.** The series is considered by contractors to be the main reference for on-site construction. Some of the parts in this document may supersede SANS 058. Other documents in the series are:
  SANS 1200 DB Earthworks (pipe trenches)
  SANS 1200 LB Bedding (pipes)

SANS 1200 LC	Cable ducts
SANS 1200 LD	Sewers

Specifying clay pipes

Consider the following points when specifying clay pipes:

- They should comply with SANS 559.
- Sewerage and drainage pipes must be specified with flexible mechanical joints.
- Rubber rings must comply with SANS 974.
- Always refer to the code of practice.
- If the pipeline is to be used for chemical effluent, additional precautions may be necessary.

Joints

A caulking material, similar to cement mortar, was initially used for joints. However, this method was discontinued because:

- the joint was so rigid and stiff that it did not allow for ground movement;
- bacteria in the sewerage attack cement-based mortars causing expansion and subsequent cracking; and
- the caulking operation required skilled pipe-layers.

The problems associated with caulked joints resulted in the design of mechanically flexible joints, of which four types are currently in use:

- rolling rubber ring;
- elastomer spigot and socket;
- sleeve and elastomer spigot; and
- plain-ended pipes with jointing sleeve and gasket.

Vitrified clay pipe applications

These pipes, buried underground, are predominantly used to convey foul wastes. Their popularity stems from the fact that they are impervious to attacks from bacteria located in sewerage and aggressive ground conditions, and perform particularly well when subject to abrasion. Vitrified clay pipes tend to stand the test of time as they do not deteriorate and can last for decades.

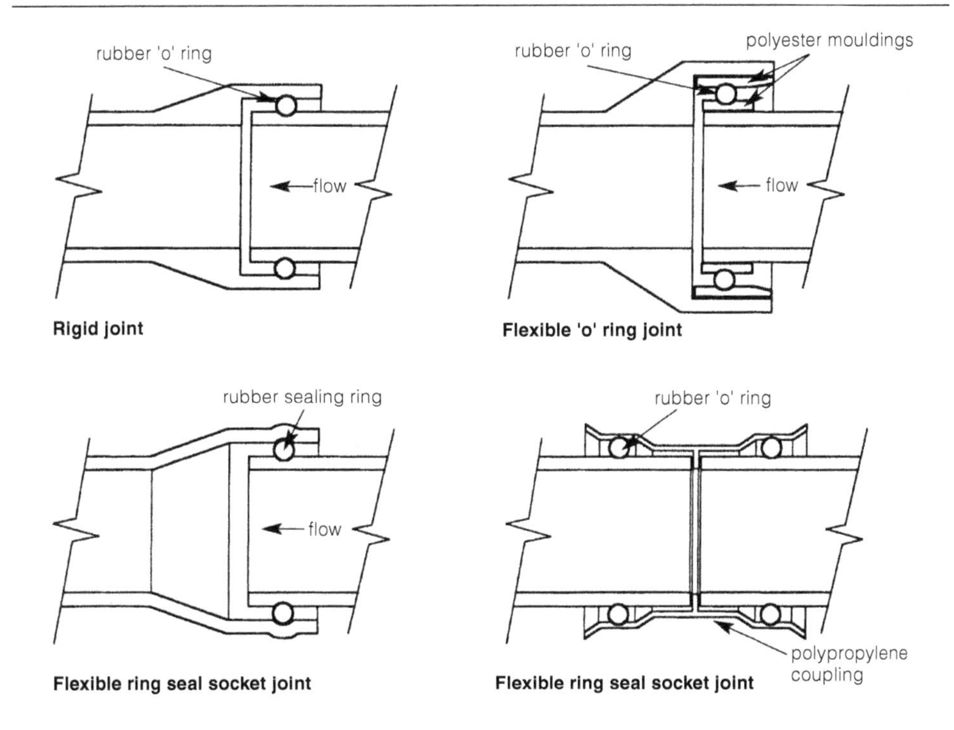

Fig 11.1 Pipe jointing

11.3.2 Drains

When designing some drainage systems ensure gradients for drainage pipes are laid to a maximum of 1 in 6 and a minimum of 1 in 60. The drainage requirements according to local authority regulations must also be adhered to.

Do you remember what 1 in 60 means? One unit in the vertical direction and 60 units in the horizontal direction. Now convert this to either a fraction or percentage.

Activity 11.1

A sewer pipe laid at a gradient of 1:60 connects two manholes. If the invert level of manhole 'A' is 100.00 and the pipe is laid sloping down towards manhole 'B', calculate what the invert level of the pipe will be at manhole 'B' if they are 30 m apart.

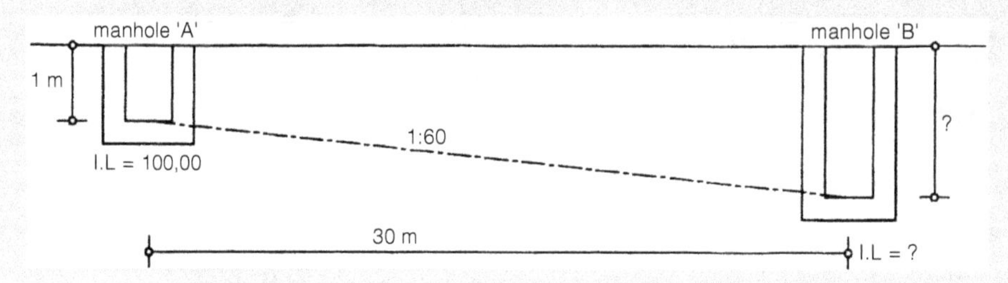

Fig 11.2 Calculation illustration

Solution

If the pipe is sloping **down**, the level at manhole 'B' will be **less** than that at manhole 'A'. The difference in level caused by the gradient must be calculated and then subtracted from the level at manhole 'A'.

To determine the difference in level, you can either retain the gradient as a fraction or convert it to a percentage and then multiply it by the distance between the two manholes.

	Option 1 (fraction)	Option 2 (%)
Difference in gradient	1:60 = 0.017	1:60 × 100 = 1.667%
Difference in elevation	0.017 × 30 m = 0.50	1.667% × 30 m = 0.50
Invert level at manhole B	100.00 – 0.50 = 99.50	100.00 – 0.50 = 99.50
Can you now determine the depth of manhole 'B'?		

Minimum cover

Drainage should have a minimum cover of 800 mm. Where it is necessary to have the pipe shallower than the minimum, it should be encased (surrounded) by 150 mm concrete. This is a decision which the 'engineer' must take depending on various conditions on site.

Why do you think it is so important to have a minimum cover on pipes?

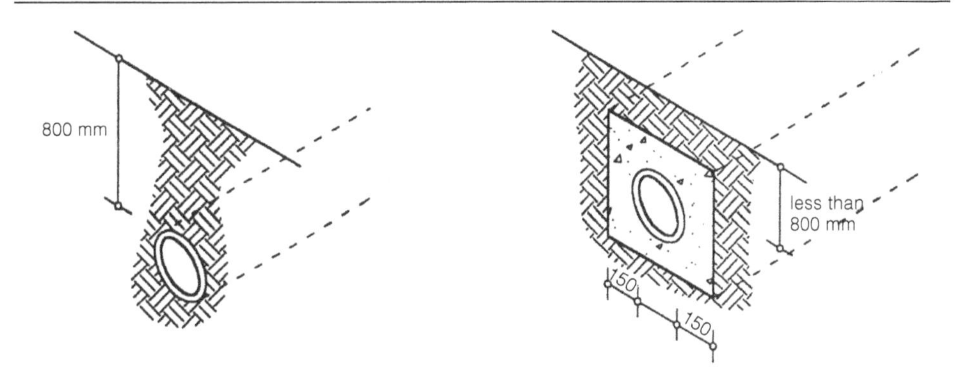

Fig 11.3 Minimum pipe cover

Access

All drains must be accessible, particularly for cleaning and for inspections. Normally manholes or inspection chambers will provide this function, but cleaning eyes or access eyes will serve the same purpose.

Flexibility

Imagine what will happen if the ground moves but the pipes remain rigid? To avoid cracking, jointing is used for all drains. Where a pipeline with flexible joints is connected to a rigid structure – for example, a manhole or an inspection chamber – differential movement is predictable. To accommodate this movement, use shorter pipes.

Anchor blocks

Anchor blocks (normally made of mass concrete) are used to prevent pipes from moving and are usually placed at bends or where the slope of the pipe is greater than 1 in 6. Anchor blocks should be made of concrete, at least 300 mm wide, and embedded into the sides and bottom of the trench to a depth of 150 mm. Remember to use short pipes on the inlet and exit sides of the anchor block.

Sewerage

Sewerage systems – pipes accepting sewage from foul water drains – are usually owned and maintained by the local authority or municipality.

Minimum gradients on sewerage pipes are determined by the minimum permissible full bore velocity. Under normal circumstances, this is not less than 0.9 m/s. Absolute minimum velocity of 0.6m/s is allowed only in exceptional cases. The danger of using this velocity must be checked against the type of material being transported, since velocity is dependent on the particle size and specific weight of the material.

Manholes

The distances between manholes will depend on the size of the sewer and the type of cleaning equipment utilised. Manholes are generally located at:

- changes in direction and gradient;
- junctions;
- the head of each sewer; and
- intervals not exceeding 100 m for straight runs.

Manholes should be constructed to permit the minimum infiltration of ground water, as this will contribute to the flow.

Manholes are normally constructed of bricks (1 200 × 800 mm) or 1 250 mm diameter precast concrete rings. The shaft or chimney of the manhole is constructed from 600 × 600 mm brickwork or 750 mm diameter precast concrete. Steel step-irons are provided inside the manhole for access.

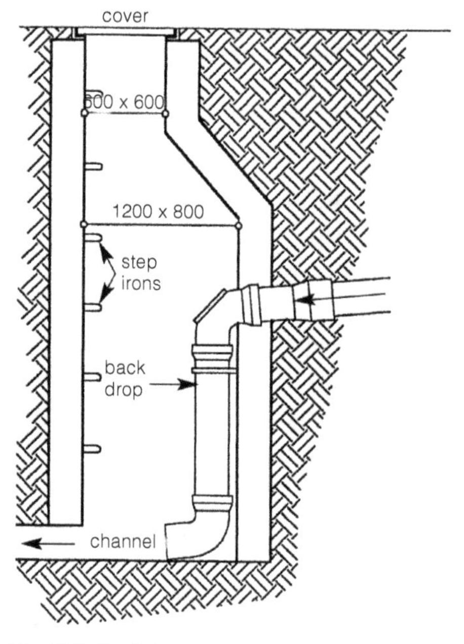

Fig 11.4 Backdrop system

The floor of the manhole consists of a half-round vitrified clay pipe (called a channel) that is secured in concrete benching. Where it is necessary to connect at different levels, a backdrop system is used, consisting of two bends and a length of pipe between them.

Installing pipes

Contractors may use the best and strongest pipes available, but if they are not installed properly, the drainage system will fail. It is important that the ground conditions are carefully checked to ensure the proper specification of equipment and materials.

When installing pipes:

- Always try to schedule pipe laying and back filling to follow as soon as possible after excavation.
- Ensure correct earth cover when back filling as overloading may occur.
- Adhere to trench widths as specified on drawings, as this determines the loading on pipes.

- Ensure correct pipe bedding.
- Shoring of excavated sides may be necessary for deep trenches.
- Keep the excavated soil away from the sides of trenches to avoid possible collapse.
- Ensure the correct alignment/grades of pipe when installing. This can be done manually (using boning rods) or mechanically (using a laser system). Boning rods are accurately measured timber members sited between two level pegs to fix pipe gradients. Laser technology uses a laser beam as a guide to determine pipe gradients.
- Use granular material to level out the trench floor.
- Handle pipes carefully to prevent the breakage of pipes and jointing.

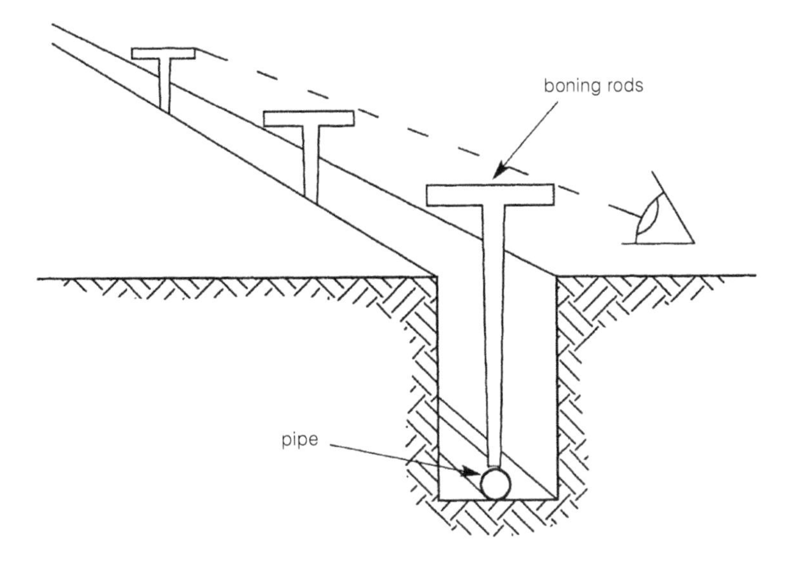

Fig 11.5 Pipe levelling using boning rods

Bedding and back filling

The bedding and back filling of pipes is fundamentally important to their performance. The type and class of bedding are usually specified in the contract documents or on the drawings. Remember to ensure that there are no hard or soft spots on the trench floor, as pipes need to lie on a firm base.

There are four types of bedding designs used in South African conditions:

- Class A bedding
- Class B bedding
- Class C bedding
- Class D bedding

The bedding conditions are in support of the type of pipe used, the ground conditions and the loading expected (on top of the pipe).

Class A bedding

- The pipes must be supported at the correct grade on softwood wedges. These wedges should be placed on either side of the socket or sleeve so that the pipe is supported at two points.
- Where flexible joints are used, a vertical construction joint using 10 mm softboard or similar material should be formed in the concrete at every second or third joint.
- Concrete is now placed from one side only and worked under the pipes to ensure that there are no voids.

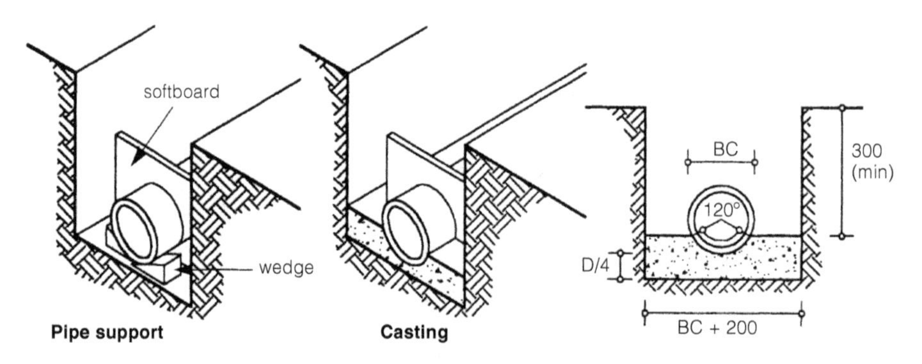

Fig 11.6 Class A bedding

Class B bedding

The initial layer of granular material must be formed and compacted with joint holes under socket or sleeve positions, to ensure that the pipe is fully supported along its barrel length. These joint holes should be approximately 50 mm deep and can be formed with softboard, polystyrene or similar material.

After laying the pipes at the correct gradient, more granular material should be placed and evenly compacted to centre the pipe.

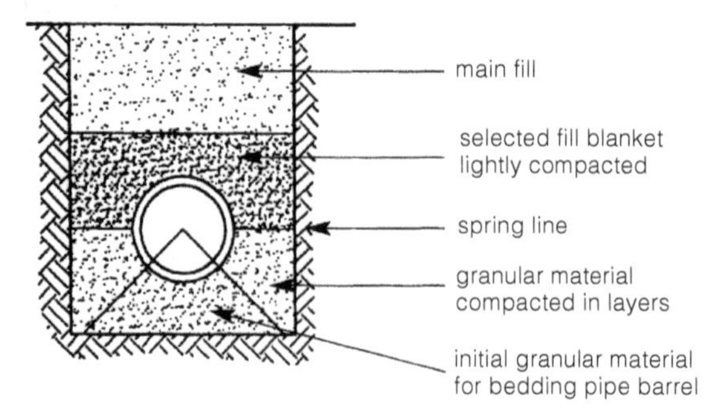

Fig 11.7 Class B bedding

Class BB bedding

This is similar to Class B, but the granular material is carried 150 mm above the top of the pipe.

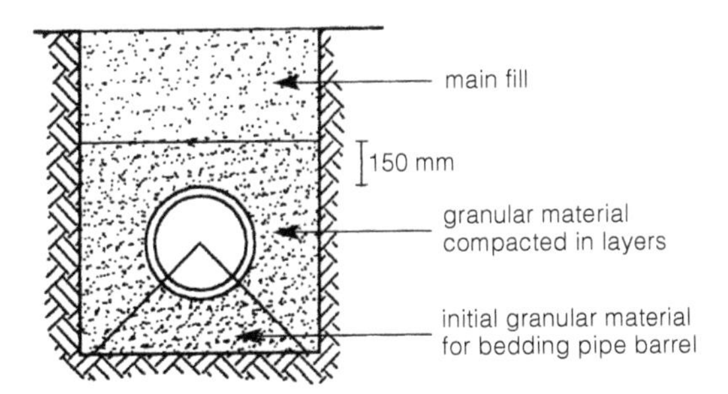

Fig 11.8 Class BB bedding

Class C bedding

Here, the granular material is only brought up to one-sixth of the pipe's diameter.

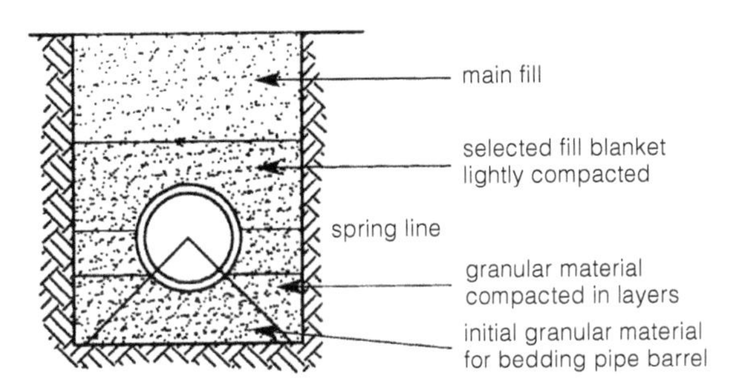

Fig 11.9 Class C bedding

Class D bedding

This is the most common bedding for pipes up to 300 mm diameter, placed in uniform, relatively dry, fine-grained soils. It is important that the joint holes are positioned under sockets and sleeves, and that the barrel of the pipe is fully supported along its length.

Back filling

Due to the brittle nature of all pipes, initial back filling should be of selected material, for example sand, placed in layers of approximately 100 mm and compacted by hand for the full width of the trench. The main filling can be placed and mechanically compacted once the initial back fill has reached a level approximately 300 mm above the pipe.

Activity 11.2

Obtain a copy of SANS 1200 and investigate the following for bedding and back filling:

- the material specification for granular material;
- the material specification for selected fill; and
- the material specification for main fill.

Make a note of the information and discuss this in class. Remember to cross-check this information with SANS 058.

Properties of clay pipes

Table 11.1 Advantages and disadvantages of clay pipes

Advantages	Disadvantages
Inexpensive due to low cost of raw material	Clay pipes are **brittle**
Reasonably strong in **compression**	Tendency to **shear** at manholes
Resistant to sewer gas, erosion, corrosion	Cannot withstand **high impact loading** or surcharge
Flexible joints allow for minor movement	Back filling under controlled conditions is required to **prevent damage**
Easy to **handle**	

11.3.3 Cast iron pipes

These pipes are generally only considered for **domestic drainage** in special circumstances, such as sites with **unstable ground**, drains with shallow inverts and drains that pass under buildings. Like clay pipes, cast iron pipes are made with a spigot and socket for rigid or flexible joints. Cast iron pipes are given a protective coating of a hot tar composition or a combination of coal solution and bitumen composition. This coating gives the pipes good protection against corrosion and reasonable durability in average ground conditions. Generally, cast iron pipes are not popular because they are expensive and it is cheaper to use PVC pipes.

11.3.4 Pitch fibre pipes

Pitch fibre pipes and fittings are made from pre-formed felted wood cellulose fibres that have been thoroughly impregnated under vacuum pressure with at least 65% (by weight) cold tar pitch or bituminous compounds. They are suitable for all forms of **domestic drainage** and because of the smooth bore, with its high flow capacity, they can generally be laid to **lower gradients** than most other materials.

Diameters range from 50 to 225 mm with a general length of between 2 400 and 3 000 mm.

11.3.5 uPVC pipes

uPVC is commonly referred to as PVC. These pipes and fittings are made from polyvinyl chloride and additives needed to manufacture the polymer and produce a **sound, durable pipe**. The pipes are obtainable with socket joints for either a solvent welded joint or a ring sealed joint. Like pitch fibre pipes, PVC pipes have a smooth bore, are light and easy to handle, come in long lengths (reducing the number of joints required) and they can be jointed and laid in all weathers.

11.3.6 Concrete pipes

Concrete pipes are made of a mixture of fine aggregate and Portland cement that is poured over a wire cage and compacted in steel forms.

Non-pressure pipes are used for **low or non-pressure applications** (e.g. storm water or sewerage) and are manufactured in different sizes and classes. Pipe design is influenced by the depth and width of the trench, back fill material and loading. Non-pressure pipes are classified according to their **proof load** and **ultimate load**. Proof load can be defined as the load that a pipe can sustain without significant cracking. Cracking should not exceed 0.25 mm in width over a length of 300 mm or more when tested. Ultimate load refers to the maximum vertical load the pipe will support when tested.

Non-pressure pipes are guided by **SANS 677**: Standard specification for concrete non-pressure pipes.

The table below indicates load classifications for non-pressure pipes.

Table 11.2 Load classifications for non-pressure pipes

Class of pipe	Proof load	Ultimate load
25D	25D (D in metres)	31.25D
50D	50D (D in metres)	62.50D
75D	75D (D in metres)	93.75D
100D	100D (D in metres)	125.00D

EXAMPLE 11.1

A 450 mm nominal diameter 50 D pipe is capable of sustaining:
Proof load = (50.00 × 0.45 m) = 22.50 kN/m
Ultimate load = (62.50 × 0.45 m) = 28.125 kN/m

Pressure pipes are used in pipelines that carry liquids under pressure (e.g. water mains). They are manufactured in different sizes and classed according to pressure. The class of a pressure pipe is defined as the factory test pressure that the pipe can sustain for two minutes without leakage.

The table below indicates pressure classes for pressure pipes:

Table 11.3 Classes of pressure pipes

Class of pipe	Test pressure in bars	Kilopascals
T2	2	200
T4	4	400
T6	6	600
T8	8	800
T10	10	1 000

Pressure pipes are guided by **SANS 676**: Standard specifications for reinforced concrete pressure pipes.

Both SANS 676 and SANS 677 cover:
- materials used;
- physical properties of the pipe (dimensions, load requirements, tests, etc);
- the standard marking of pipes;
- methods of sampling for inspection, testing and compliance with specifications; and
- methods of testing pipes at the factory.

Activity 11.3

To enhance your understanding of concrete pipes, approach concrete pipe manufacturers or other sources of information to find out how concrete pipes are made, the processes followed and what specifications are applied.

Pipe loads

Every pipeline is subject to forces that cause stresses in the pipe wall. These can be divided into primary and secondary loads.

Of the two loads, secondary loads can cause considerable damage through differential movement of the pipeline and should not be underestimated in the design process.

Table 11.4 Differences between primary and secondary loads

Primary loads	Secondary loads
Can be calculated	Not easily calculated
Mass of earth above fill	Change in clay volume when moisture content changes
Traffic loads	Pressures due to growth of tree roots
Internal pressure loading	Settlement of foundations
Mass of pipe	Thermal and moisture changes in pipe material
Mass of water	Restraint caused by manholes

Jointing

There are essentially two types of pipe joints: those with a rubber sealing gasket and those that do not perform a sealing function. Typical joints that do not provide a sealing function are either Ogee-ended (self-centring) or butt-ended joints.

Joints that include a sealing gasket are 'spigot and socket' and 'in the wall' joints for larger diameter pipes. Sealing gaskets can either be of the rolling rubber ring type or the confined rig (sliding joint) type.

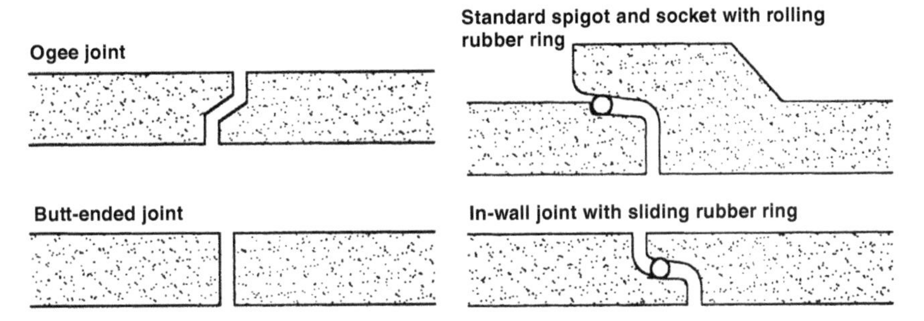

Fig 11.10 Straight joints (non-sealing)

Fig 11.11 Joints with sealing gasket

Installing concrete pipes

Bedding plays a significant role in the installation of concrete pipes, as it transfers the vertical load on the pipe to the soil foundations. It is important that the material used for bedding and back filling conforms to the specifications prescribed by the code of practice.

Class A Concrete bedding

Once the line and grade have been accepted, concrete bedding is cast around the pipe. Care must be taken to ensure that the concrete is well compacted under the pipe. The concrete must have a 28-day cube strength of not less than 20 MPa. Only once the concrete has reached a strength of 10–15 MPa is back filling recommended.

Class B Granular bedding
Granular bedding material should be thoroughly compacted under and around the sides of the pipe.

Class D bedding
No special precautions are required for this class of bedding, except that the subgrade must fully support the pipe along its length, including the joints or sockets with a diameter greater than the barrel of the pipe.

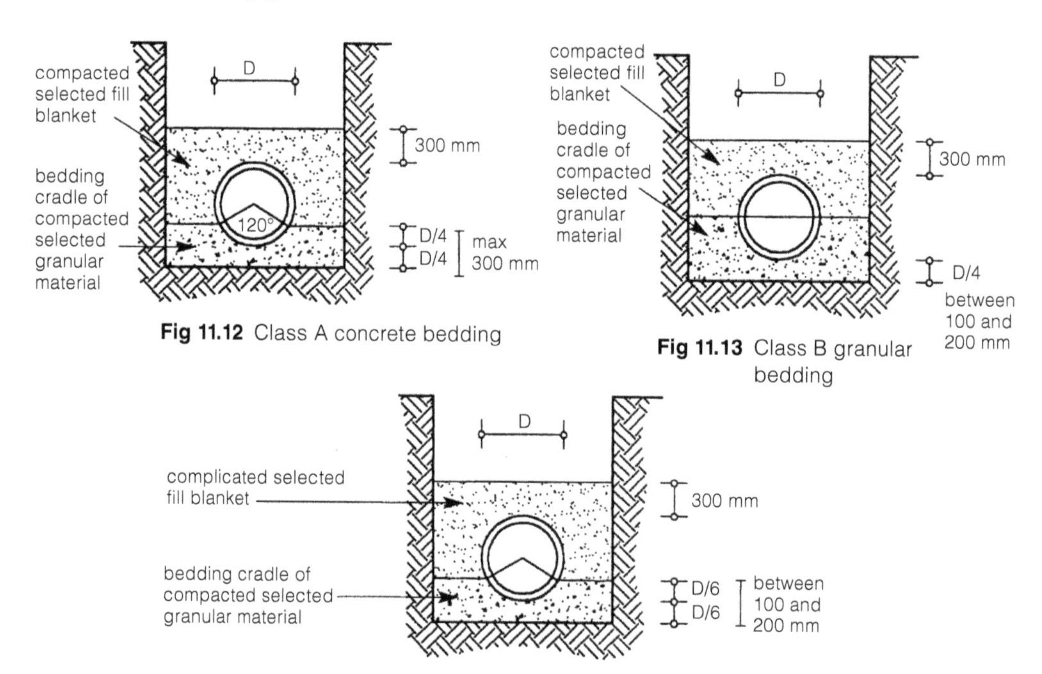

Fig 11.12 Class A concrete bedding

Fig 11.13 Class B granular bedding

Fig 11.14 Class D bedding

Properties of concrete pipes

- Due to concrete's density, concrete pipes are relatively strong.
- Good corrosion-resistant properties.
- The pipes are made in two classes, allowing them to be used in pressure or non-pressure situations.

Methods of jointing

It is essential that pipes are joined correctly to ensure proper sealing and strength at pipe joints. Here are some of the rules to be followed when joining pipes:

- Ensure that the surfaces of the joint and the rubber ring are clean.
- Align the pipes correctly before jointing.

- Lubricate the rubber ring and grease the socket before placing in the groove.
- The spigot end of the pipe to be laid should be drawn up so that it just enters the socket of the previously laid pipe.
- For pipes smaller than 600 mm (on a non-granular bedding) use a wooden block behind the socket end of the pipe and a vertical bar as a lever to move the pipe into position.
- Use a sling at the centre of gravity for heavier pipes to align the pipe.
- Thereafter use a pulling device or turfor to pull the pipe into position.
- Pipes with socket joints are laid against the flow – they start at the lower end and work towards the upper end.
- The collar of the socket is laid in a prepared hollow in the bedding and the ball is centralised.

11.4 Surface drainage (rainwater)

You will find some form of rainwater channeling on all building structures. Semi-circle or box-shaped gutters are fitted to buildings to collect rainwater that falls on the roof and conducts the flow into downpipes. The rainwater pipe is terminated at its lowest point by a rainwater shoe for discharge into a surface water drain or a trap gulley. From here, the flow is directed into further channels on the property or into the formal stormwater drainage system usually located in the road.

The pipes used for domestic rainwater installations are made of asbestos or unplasticised polyvinyl chloride (uPVC). The advantages of uPVC include:

- easier jointing;
- no gutter bolts are required;
- the joint is self-sealing (the pipe has a built-in rubber seal at the joint);
- there is no corrosion;
- decoration (painting) is usually not required as the pipe is often used underground;
- breakages are reduced; and
- better flow properties enable smaller sections and gentler falls.

Semi-circle and box-shaped uPVC pipes are usually between 75 and 100 mm, and come in lengths from 1.8 to 3.6 m.

Rainwater drainage installation is needed to collect discharge from roofs and paved areas, and convey it to a suitable drainage system. We will look at road drainage and drainage collected from roof water.

It is important that rainwater is directed away from road surfaces as quickly as possible. The reason for this is that surface water affects the traction between the vehicle tyres and the road surface that could cause skidding. Also, if water is allowed to penetrate the structural layers of the road, this will result in them softening and will accelerate deterioration.

Water spray is another problem resulting from rainwater on road surfaces. Imagine driving behind a truck in the rain. Visibility is poor and can cause accidents.

The crossfall and the gradient of a road govern surface drainage. The flatter the gradient and the shallower the cross fall (which is near to horizontal in both cases), the slower the water flows across the surface. You will find that the specifications used for the cross fall design of a road is on average 2.5% to either side of or across the entire surface of the road. This ensures that water runs off as quickly as possible. Several methods are used to collect and channel water from roadways (see Chapter 3).

Activity 11.4

Look at the roads you travel on and notice how water is collected. Make a note of this or sketch the arrangement you see so that it can be discussed in class or compared with other drainage methods.

One method of collecting water is by using kerbing with adjacent channels to conduct the water to a catch pit, from where it is drained into storm water pipes. Alternatively, water can be drained through flat or V channels.

Water can also flow completely off the road and be collected in channels or drains, that conduct it to grid-top manholes and into stormwater pipes.

Problems in collecting and disposing of surface water often arise when there is high ground alongside a road or the road has been constructed in a cutting. However, this may be solved by constructing a drain (known as a catchwater berm) along the high ground side and, at intervals, conducting the water down chutes or pipes to a drain. There should be a channel to collect this water along the toe.

The design of pipe systems is a specialist field and aspects discussed under this heading will be covered in later courses. Discussions here are presented as an introduction to the topic.

Surface drainage comprises two main aspects:

- The volume of water arriving at a ditch or culvert relates to the **hydrological study**.
- The design of the facility to handle water is referred to as the **hydraulic study**.

11.4.1 Hydrological study

A fundamental feature of rainfall is that the intensity of a given storm is inversely proportional to the length of the storm (although this can be severely affected by, for example, hurricanes). Hardened surfaces, like car parks and roofs, all contribute to water flow that collects and should be disposed of adequately. Usually the problem is more critical in urban areas than in rural areas.

Flooding in Mozambique in February/March 2000 resulted in widespread devastation. To combat this situation we need to design facilities or services (like stormwater pipes of various diameters) and locate them in various positions to allow for the movement of water with minimal damage.

It is important to know what happens to water during a storm, and this is referred to as the **hydrological study**. Internationally, storm duration is used as the benchmark for design purposes. There is the assumption that the maximum discharge at any point in the drainage system occurs when the entire catchment area is contributing to the flow. The rainfall intensity producing this flow is the average rate of rainfall that can be expected to fall in the time required for a raindrop to travel from the most remote part of the catchment area to the point under consideration. The time taken for the raindrop to make this trip is called the **time of concentration** (duration). This period is influenced by factors such as the slope of the ground, the nature of the surface and the length of the flow path. All of these variables are usually incorporated in a nomogram, from which the time of concentration (T) can be determined for the catchment area. From this, the corresponding intensity (I) for a given storm frequency (how often a storm occurs) can be determined.

This **rational formula** has been developed to calculate the runoff for small catchment areas:

$q = A \times I \times P$

where

q = run off in cubic metres per second

A = catchment area in square metres

l = rainfall intensity in metres per second

P = runoff coefficient

The coefficient 'P' gives the proportion of the runoff that will reach a certain point after allowing for soakage and evaporation. The factors governing this include:

- The type of soil in terms of infiltration – sandy soils will infiltrate much better than claysoils.

- The nature and extent of the vegetation – the more sparsely grown the area, the quicker the runoff; the more overgrown and vegetated the area, the slower the runoff.
- Length and steepness of slopes – steeper slopes mean faster runoff; flatter slopes mean slower runoff.
- The size and shape of the catchment area.
- Atmospheric temperature – the higher the temperature, the greater the evaporation rate.

Small catchments vary from a few square metres to about 15 square kilometers. Typical small catchment problems are:

- the sizing of gutters and downpipes for drainage roofs, elevated roads or similar structures;
- the sizing of gutters and storm water drains for street drainage;
- storm water drainage of airports, sports fields, parks and gardens;
- the design of outfall drains, highways and railway culverts, and the cross drainage of channels;
- the sizing of waterways beneath bridges;
- the design of agricultural terraces and grassed spillways;
- the sizing of spillways for small catchment dams;
- determining freeboard allowance for dams without spillways; and
- providing technical information for the settling of insurance claims or for the handling of legal issues arising from storm or flood damage.

To obtain information from a hydrological study, it is important that an engineer or technician makes a thorough investigation of the site and its condition, in addition to the climatic conditions. The weather bureaus at airports and government departments usually have information about rainfall figures, temperatures and other climatic conditions. Talking to the local population will also provide valuable information.

11.4.2 The hydraulic study

After determining the quantity of runoff, the focus becomes the hydraulic design of the pipe, channel or culvert to carry it. In an open channel, flow takes place under gravity,

The **Manning formula** is often used to determine channel flow and pipe flow:

$$V = \frac{1}{n} \times \frac{r^2}{3} \times \frac{s1}{2}$$

Remember that $q = V \times A$ and $r = \frac{A}{p}$

where

- n = Manning roughness coefficient, which varies from 0.009 to 0.035 depending on the material
- r = hydraulic radius ($\frac{A}{p}$)
- s = the slope
- v = velocity in metres per second (inside the pipe)
- q = quantity in cubic metres per second, which is also equal to $v \times A$
- A = the area
- p = the length of the wetted perimeter

11.5 Pipe and gutter sizing

The size of the gutters and downpipes to effectively cater for the discharge from a roof will depend on:

- the area of roof to be drained;
- the anticipated intensity of rainfall;
- the material used for the gutters and downpipes;
- the fall within the gutter; and
- the number, size and position of outlooks.

Building regulations usually indicate the required size as per the area and factors mentioned above.

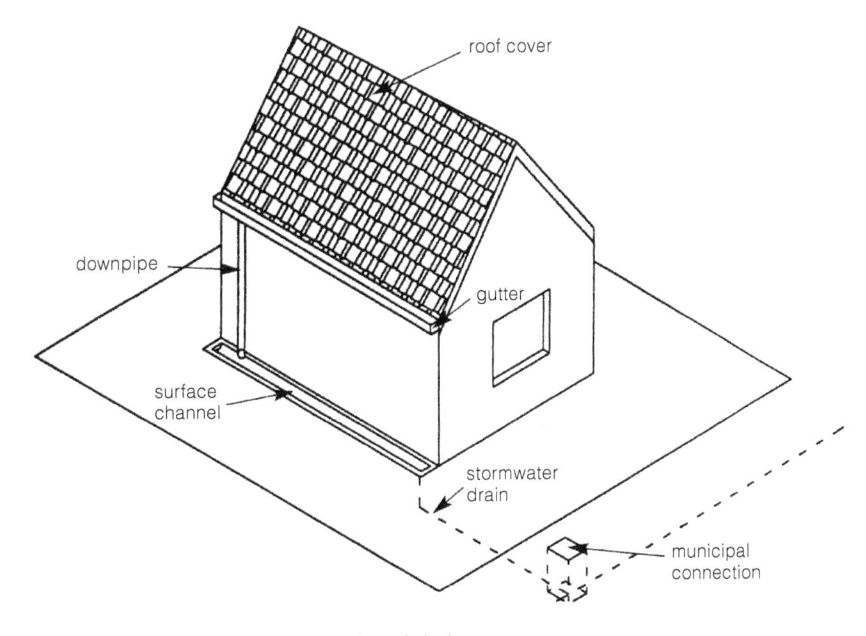

Fig 11.15 Rainwater pipework and drainage

We will not be covering drainage elements within a building – for example, urinals, toilets, hand basins, baths, etc. – as these functions relate specifically to building and building construction and do not form part of this Civil Engineering study programme.

11.5.1 Pipe testing

A requirement stipulated by law, in the Building Regulation Act of South Africa, is that all pipes must be tested for water tightness. Several factors must be taken into consideration when testing pipes.

- The local authority can determine the type of test to be applied.
- The test must be for water tightness. It can be argued that only a water test can fulfil this function although many local authorities use a smoke test.
- Testing must be carried out after back filling. It is therefore in the contractor's interest to test the drains before the back filling is carried out, since the detection and repair of any failure discovered thereafter can be time consuming and costly.

11.5.2 Types of tests

Four methods can be used to test drains:

- **Water test.** This involves filling the drain with water under pressure and observing if any water escapes.
- **Smoke test.** Smoke is pumped into the pipes and the float on the smoke machine is checked for any fall in pressure.
- **Air tests.** This is not a particularly conclusive test, but it is sometimes used in special circumstances, such as large diameter pipes, where a large quantity of water may be required. If failure is indicated by an air test, the drain should be re-tested using the more reliable water test. In South Africa, the water test is the test most often applied on construction sites.
- **Visual inspection.** This test is performed on pipes that have large enough diameters to allow a person to crawl into them.

Smoke tests mostly check for escaping smoke while air tests involve pressure drop observations.

In general, drain testing should be carried out between manholes. Manholes should be tested separately and short branches of less than 6 m should be tested along with the main drain to which they are connected. Long branches should be tested in the same manner as the main drain.

11.5.3 Ventilation

To prevent foul air from soil and drains becoming a nuisance, all drains should be vented with a flow of air. A ventilating pipe should be provided at or near the head of each main drain and any branch drain exceeding 10 m in length. The ventilating pipe can be a separate pipe or a soil discharge stack pipe. A vertical stack pipe is often bolted to the wall to vent manholes.

Please ensure that when inspecting manholes, both the upstream and downstream manholes from the one you are inspecting are opened. Methane gas commonly builds up inside sewerage systems and if exposed to this for a short time, it can be lethal – so don't immediately climb into a sewer manhole of an active system **and always ensure you wear a harness.**

11.6 Private sewers

A sewer can be defined as a means of conveying sewage and stormwater collected from the drains, below the ground, to the final disposal point. If a sewer is owned and maintained by the local authority, it is generally called a public sewer. A sewer owned by a single person or a group of people and maintained by them is classed as a private sewer.

When planning the connection of houses to the main or public sewer, one method is to consider each dwelling in isolation. However, important economies in design can be achieved by using a private sewer.

A number of houses can be connected to a single (private) sewer that, in turn, is connected to a public sewer. Depending on the number of houses connected to the private sewer and the distance from the public sewer, the savings are possible in the following areas:

- total length of drainpipes required;
- the number of connections to the public sewer;
- the number of openings in the road; and
- the number of inspection chambers.

11.7 Pipe connections

Building regulations require that all sewer connections are made so that the in-coming drains or private sewers join to the main sewer obliquely in the direction of flow and that the connection will remain water tight and satisfactory under all working conditions. Normally, sewer connections are made by the local authority or under their direction or supervision. The method of connection will depend on several factors:

- the relative sizes of the sewer and the connecting drain or private sewer;
- relative invert levels;
- the position of nearest inspection chamber on the sewer run;
- whether the sewer is existing or being laid concurrently with the drains or private sewers;
- the type of joints or junctions built into the existing sewer; and
- the shortest and most practicable route.

If the public sewer is of a small diameter, less than 150 mm, the practical method is to remove two or three lengths and replace them with an oblique junction to receive the incoming drain.

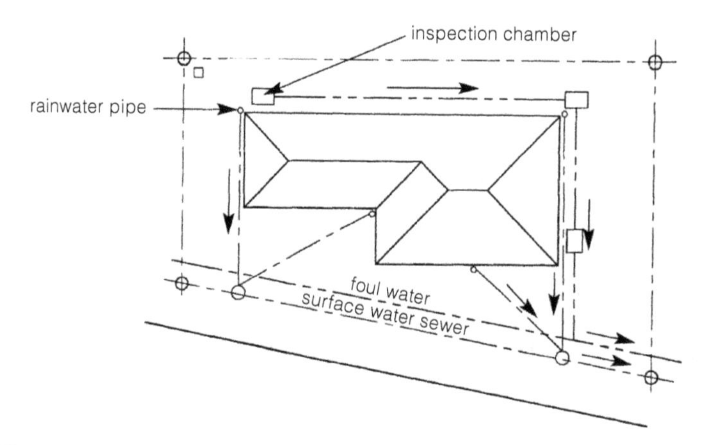

Fig 11.16 Private sewer arrangement

Connections to inspection chambers or manholes, whether new or existing, can take several forms depending mainly upon the different invert levels. If the invert levels of the sewer and incoming drains are similar, the connection can be made in the conventional way using an oblique branch channel. Where there is a difference in invert levels, the following can be considered:

- a ramp in the benching (sloped areas at the bottom of manholes shaped by mortar) within the expecting inspection chamber;
- a backdrop manhole or inspection chamber; or
- increasing the gradient of the branch drain.

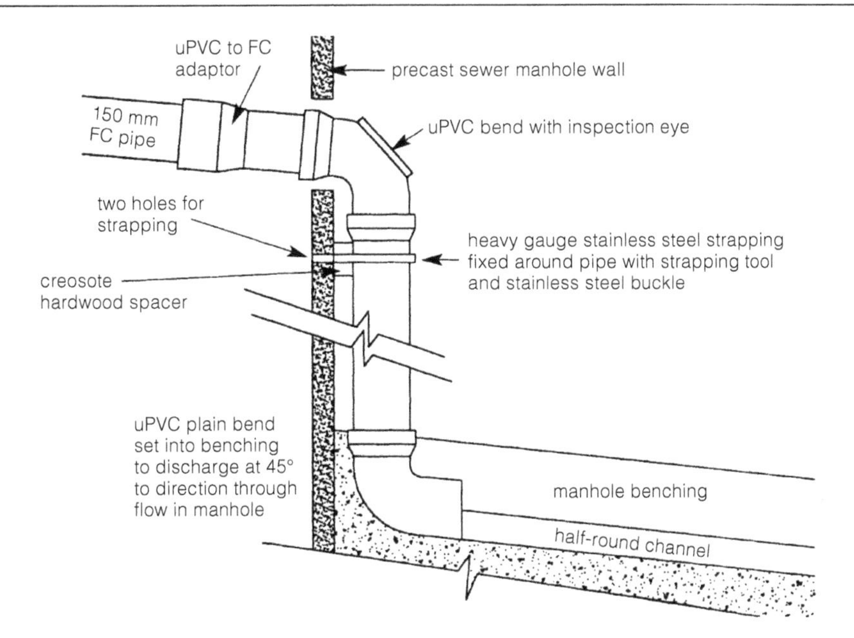

Fig 11.17 Ramp and benching for backdrop manhole

Changing the gradient of the incoming drain to bring its invert level in line with that of the sewer requires careful consideration and design. Although simple in concept, the gradient must be such that a self-cleansing velocity is maintained and building regulations are not contravened.

Connections of small diameter drains to a large diameter sewer can be made by any of the methods described above or by using a saddle connection. A saddle is a short socketed pipe with a flange or saddle curved to match the outer profile of the sewer pipe. To make the connection, a hole must be cut in the upper part of the sewer to receive the saddle, ensuring little or no debris falls into the sewer. A small pilot hole is usually cut and then enlarged to the required diameter, The saddle connection is then bedded onto the sewer pipe with a cement mortar and the whole connection is surrounded by a minimum of 150 mm of mass concrete.

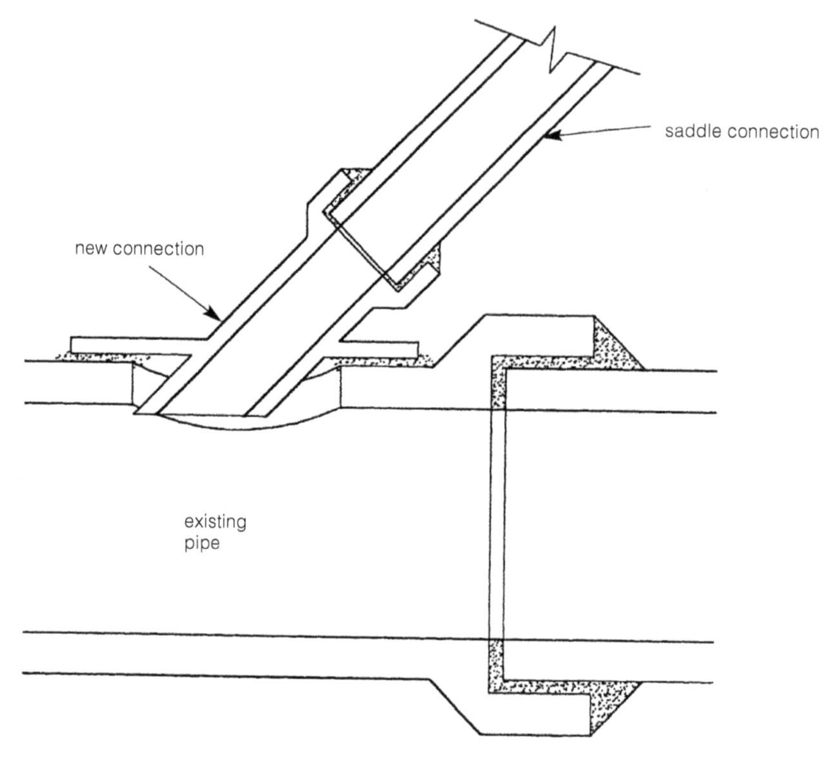

Fig 11.18 Saddle connection

11.8 Soak-aways

A soak-away is a pit dug in permeable ground that receives rain water or sewerage water. It is constructed in such a way that the water can percolate into the surrounding subsoil. To function correctly and efficiently, a soak-away must be designed taking into account the following factors:

- the permeability or rate of dispersion of the subsoil;
- the area to be drained;
- the storage capacity of the soak-away needed to accept a sudden inflow of water;
- the local authority requirements regarding method of construction and siting in relation to buildings; and
- the depth of the water table.

Before any soak-away is designed or constructed, the local authorities should be contacted to obtain permission and ascertain specific requirements. Some authorities will allow soak-aways to be used as outfalls to subsoil drainage schemes or to receive effluent from a small sewage plant.

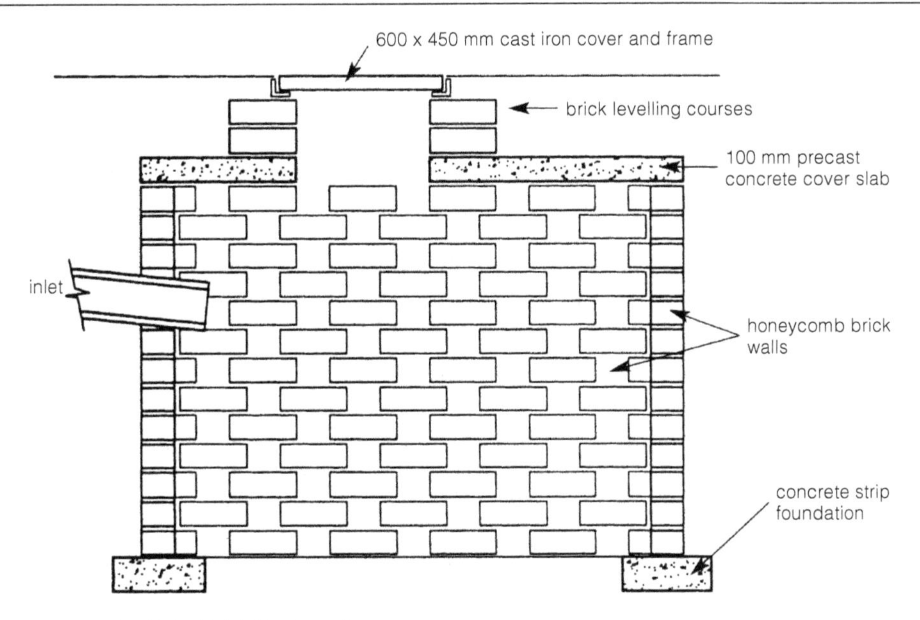

Fig 11.19 Brick-lined soakaway

The rate at which water percolates into the ground depends on the type of soil where it collects. Generally, clay soils are unacceptable for soak-away construction, whereas sand and gravel are usually satisfactory.

In South Africa, soak-aways can be used in rural and urban developments, depending on the availability, position and size of the municipal or public sewer system.

11.9 Principles of good drainage

- The materials used for the pipes should have adequate strength and durability.
- The diameter of the drainpipe should be as small as is practicable. Pipes that are put in soil should have a minimum diameter of approximately 100 mm. For surface water pipes – for example, gutters – the minimum diameter is 75 mm.
- Every part of a drain should be accessible for inspection and cleaning.
- Drainpipes should be laid in straight runs are far as possible. For obvious reasons, the more curves, corners or joints in the pipe, the more difficult it will be to clean.
- Drains must be laid to a gradient that will render them self-cleansing. The fall or gradient should be calculated according to the rate of flow, velocity required and the diameter of the drain.
- Inspection chambers or manholes should be placed at points of change. The direction size of pipe, junctions and the gradient may hinder a drain from being readily cleansed.

- Inspection chambers or manholes must also be placed within 12.5 m of a drain junction. The maximum distance allowed between manholes is 90 m.
- Drain junctions must be arranged so that the incoming drain joins at an oblique angle in the direction of the main flow.
- Try to avoid laying drains under buildings. If this is unavoidable, they must be protected to ensure water tightness and to prevent damage to the building or to the pipes. Standard protection methods include:
  - encasing the drain with 150 mm minimum of mass concrete; and
  - using cast iron pipes under the building.
- Drains that are within 1.0 m of the foundations for the walls of buildings and below the foundation level must be backfilled with concrete up to the level of the underside of the foundations.
- Where possible, the top of the drain should be a minimum of 450 mm below the ground surface to avoid possible damage from ground movement and traffic.

11.10 Sub-surface drainage

Subsoil drains are pipes that are placed below the soil surface to control the water level. They are used in many different applications, but mainly for roads and agriculture. There are several ways of constructing subsoil drains using precise specifications, but generally they consist of a piping system, a geotextile fabric acting as a filter for the fine material and coarse aggregate.

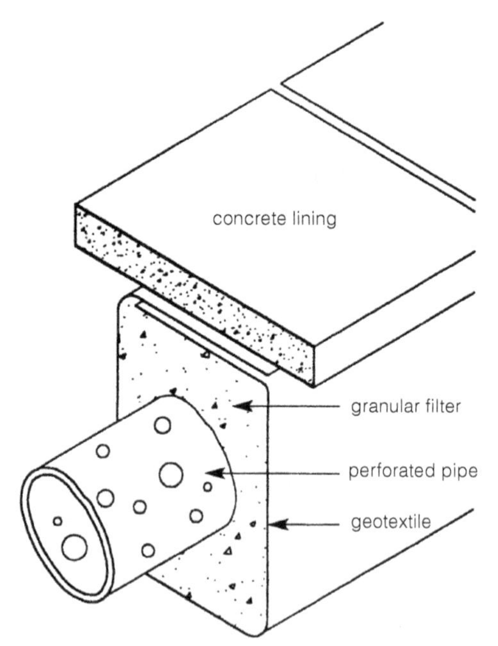

Fig 11.20 Subsoil drain

Water can reach a road foundation by infiltration from the surface, through natural seepage from the high side of the road, or through capillary rise from the water table.

By installing subsoil drains, most of this water can be controlled. If the drains are installed correctly, they can remain effective for a long time.

Interceptor drains are situated on the high side of and parallel to the road. Their purpose is to intercept seepage water and to drain any water that may enter through the surface and drain out of the base course.

Ground water drains are used for lowering the water table and are situated on either side of the road, running parallel to the alignment.

11.11 Channels

Channels are conduits typically used to alter the flow path of existing watercourse, the practice of which should be subject to environmental impact studies. Again, this is a specialised field, but mention is made here of various scour prevention tools – for example, gabions and reno mattresses.

Gabions can also be used in retained structures. These are basically wire baskets manually filled with large stones and placed in positions where they can provide protection against erosion.

11.11.1 Open drainage channels

Advantages of these types of channels include:

- low construction costs;
- large discharge; and
- storage capacities and multiple uses in recreational, aesthetic and sociological fields.

The disadvantages include:

- the space that is occupied;
- the degree of maintenance required; and
- the possible abuse of unscrupulous people.

Ideal channels are those that have been created naturally over a long time and reached some degree of stability. Where artificial channels are created, they should correspond to normal drainage patterns. There are two main types of open channels: natural and artificial.

The advantages of using natural channels include their low cost and freedom from maintenance. In their natural state, they will have reached some degree of stability.

The disadvantages are that, having formed in circumstances different from those created by urban development, they may suffer from scour and sedimentation, and require some protection treatment.

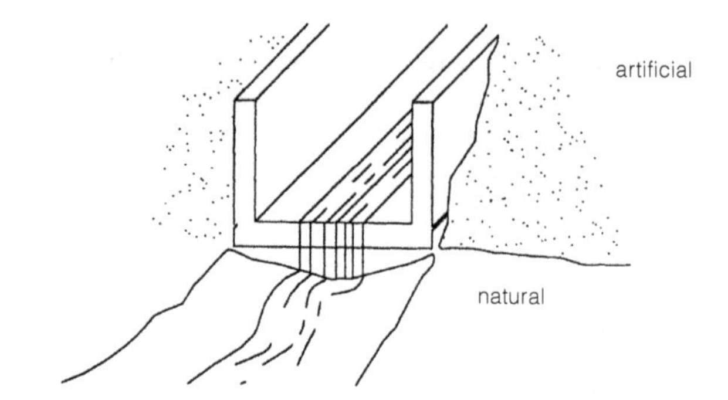

Fig 11.21 Open channels

There are several types of artificial channels of which the grassed and lined types are more common. Grassed channels have many attractions, including low flow velocities, sociological and aesthetic benefits, and low construction costs. A disadvantage is the tendency to scour at full flow.

Channels are lined for hydraulic or topographical reasons, or because of space restrictions. Lining channels increases the flow and discharge velocity. This, in turn, can compound drainage problems downstream.

Channels should only be lined when there is no other reasonable solution to drainage problems.

11.12 Culverts

A **culvert** is defined as any conduit that conveys water through an embankment. Culverts are divided into two main types: flexible and rigid.

Flexible culverts are usually either thin-walled steel pipes or galvanised corrugated sections. Rigid culverts often comprise reinforced concrete, precast units, cast iron or vitrified clay. Most of the larger culverts have concrete wing walls to aid the smooth movement of water from the river to the culverts. This not only increases the design capacity of the culvert, but also protects the surrounding soil from excessive erosion. End walls and aprons are provided for the same reason.

The location of a culvert is very important. A culvert is usually placed in the natural ditch or streambed so that its alignment conforms closely to that of the original situation. If the stream tends to meander, it should be taken through the embankment and then returned to its original flow path as soon as possible.

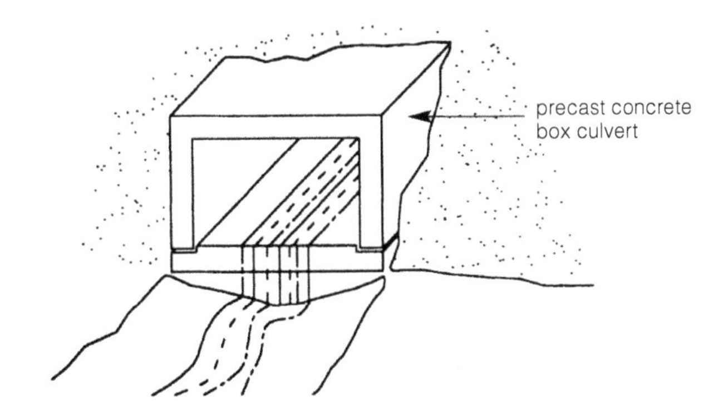

Fig 11.22 Culvert

The slope of a culvert should conform to the natural grade of the stream because this produces the least silting or scouring.

The silt-carrying capacity of the stream varies, as reducing the slope changes the square of its velocity. A minimum slope of 0.5% is recommended to prevent excess sedimentation. If the culvert slope is greater than the natural slope of the watercourse, the increased velocity may cause scouring at its exit or outlet. Several factors govern the hydraulic capacity of a culvert:

- the depth of head and tail water;
- the type of entrance. If the culvert has a poorly designed entrance, turbulence will occur at the inlet and energy that would normally be available for moving the water through the culvert dissipates. Flared or splayed entrances are more efficient than ones with straighter walls;
- the roughness of the interior walls;
- the length of the culvert; and
- the fall through the culvert.

11.13 Water supply and sewerage systems

Again, this is a specialist field and these notes should be regarded as an introduction to later water engineering, water supply and sewerage courses.

11.13.1 Water supply

Water is essential to life, and to social and economic progress, yet in South Africa there are a large number of people (in both rural and urban areas) without adequate water supplies. Adequate water supply comprises:

- reasonable access to a water source;
- the availability of a sufficient quantity of water; and
- an acceptable quality of water.

We know that our drinking water is supplied from reservoirs, but before the 'pure' water reaches our taps, it must be chemically treated to purify it. Purification takes place at large water-purification plants where chemicals are added and stringent testing is undertaken to ensure the water is fit for human consumption.

Once the water quality has satisfied the Department of Health regulations, it is transported using large pipes to the city and via smaller pipes to homes.

The demand for water supply varies according to use and seasons. For example, a higher demand is experienced during summer and this fluctuation must be catered for when designing capacities of water supplies and pipe sizes. The water flow in the pipes exerts tremendous pressure (force) and the pipelines must be able to withstand these pressures. Anchor blocks must be provided at changes of direction, to relieve the stresses on the pipeline.

Have you felt the force of water coming from your hosepipe? Now imagine the force of water in a pipe one hundred times larger. By the time the water reaches our taps, the pressure has dissipated, allowing us to use clean water for domestic, industrial or agricultural use.

Should you wish to read up more information about drainage and pipes, please access the following websites and the material (documents) contained within:

- http://www.cpc.com.sa/backfilling.html
- http://www.rocla.co.za/CMA/CMA_Pipe_and_Culverts_Installation_Manual_2003.pdf
- http://www.cma.org.za/Portals/0/Docs/Infrastructure/concrete_pipe_and_portal_culvert_handbook_edition_5_2009.pdf

11.14 Summary

The aim of this chapter was to:

- Demonstrate the importance and application of various drainage systems
- Identify drainage system components and the materials from which they are made
- Explain the testing methods used on pipelines
- Discuss the design process of a typical drainage system
- Discuss the importance of applying good drainage practice.

Self-evaluation 11

1. Complete the sentences:
   a. _______ transport water for domestic, industrial and agricultural use.
   b. Stormwater pipes are usually made of _______.
   c. The crossfall and the _______ of a road governs surface drainage.
   d. The volume of water arriving at a ditch or culvert relates to the _______ study.
   e. The _______ formula is used to determine channel and pipe flow.
   f. The _______ is a pit dug in permeable ground that receives rain water or sewerage.
   g. _______ are conduits typically used to alter the flow path of an existing watercourse.
2. State whether the following are true or false:
   a. A reservoir is a barrier of concrete or earth built across a river to create a body of water.
   b. Anchor blocks are used to prevent pipes from moving.
   c. In class C bedding, the granular material is carried 150 mm above the top of the pipe.
   d. Joints that include a sealing gasket are called 'spigot and socket' joints.
   e. Pipe testing must be carried out after back filling.
   f. Drain junctions must be arranged so that incoming drains join at an oblique angle in the direction of main flow.
   g. A culvert is a pipe placed below soil surface to control the water level.
3. Answer the following short questions:
   a. Why do you think that incoming drains should join at an oblique angle into the direction of main flow?
   b. State and describe the tests applied to pipes after they have been placed.
   c. Name the different pipe joints used in civil engineering practice.
   d. What is the difference between 'proof load' and 'ultimate load'?
   e. A sewer pipe laid at a gradient of 1:80 connects two manholes spaced 60 m apart. The invert level of the pipe at manhole B is 125.30 m and the pipe is sloping up towards manhole A. Calculate the invert level of the pipe at manhole A.
   f. If the invert level of the pipe at manhole 1 is 100.00 and the invert of the pipe at manhole 2 is 98.636 and they are spaced 45 m apart, calculate the gradient of the pipe between manhole 1 and manhole 2.

Answers to self-evaluation 11

1. a. pipes
   b. concrete
   c. gradient
   d. hydrological
   e. Manning
   f. soak-away
   g. channels
2. a. false
   b. true
   c. false
   d. true
   e. false
   f. true
   g. false
3. a. your opinion
   b. see ref 11.5.2, page 234
   c. see ref 11.3.1, page 215 and 11.3.6, page 225
   d. see ref 11.3.6, page 225
   e. 124.55 m
   f. 1:33

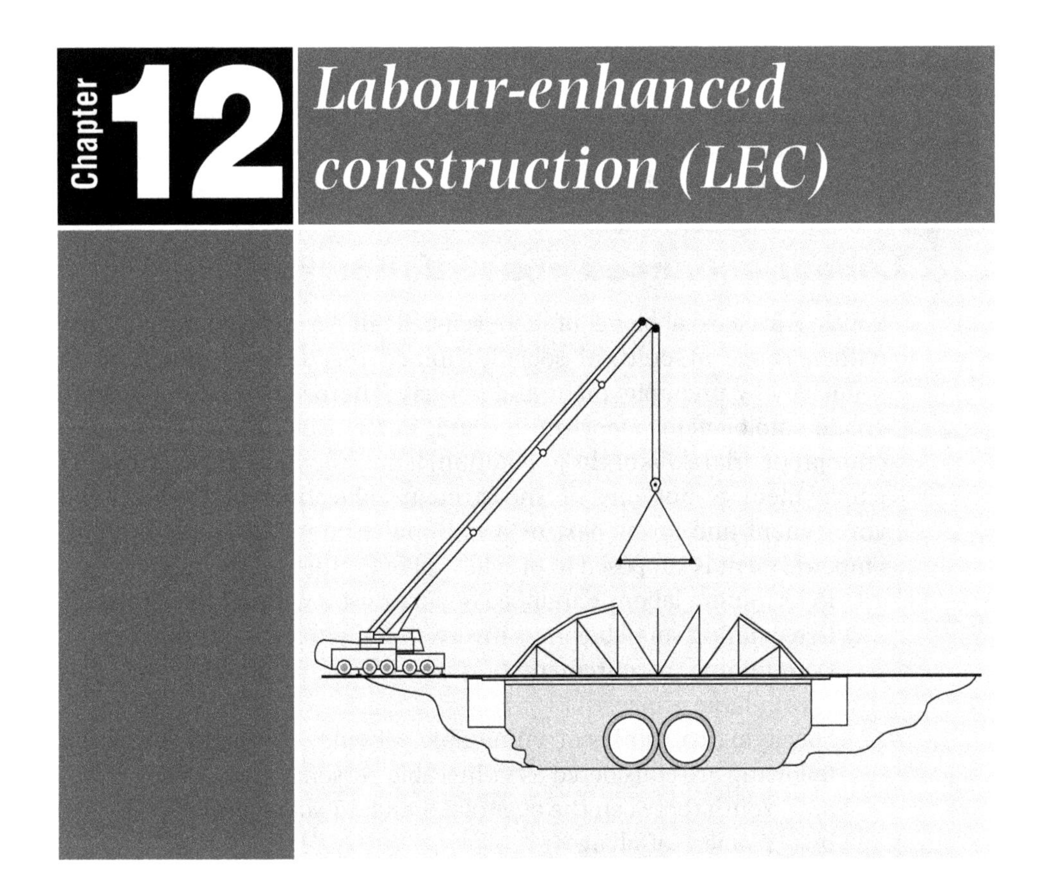

Chapter 12 Labour-enhanced construction (LEC)

Outcomes

After studying this unit, you should be able to:

- Explain the concept of labour-enhanced construction (LEC)
- Demonstrate the application of LEC to South African conditions
- Identify the objectives and principles behind LEC
- Explain the processes in the national approach to LEC.

12.1 Introduction

In this chapter we will look at some of the general issues relating to construction methods in South Africa and to the application of labour-enhanced construction in the South African context.

12.2 Labour-enhanced construction (LEC)

South Africa is a blend of a developed and a developing country, with a range of cultural societies joined together or influenced in some way by unemployment and poverty. There is therefore a need to create employment opportunities and, at the same time, develop the entrepreneurial skills of the population. It must be viewed that providing employment is but one of the strategic objectives of the national government and forms part of a national strategy of empowerment. Empowerment is supposed to provide opportunities for the following:

1. Determining labour enhanced opportunities within large projects
2. Determining small business involvement within large projects
3. Determining procurement of local labour, plant and materials within large projects.
4. Specifying categories of vulnerable persons – in South Africa the following are considered as vulnerable persons:
    a. Youth (between the ages of 18 and 35 years old)
    b. Women (of all ages)
    c. Persons with disabilities (mental or physical)
5. Establish education, training and skills transfer for the affected Labour Force
6. Developing targets such as targeted labour, local resources and training of labour within large projects.

The national government has put in place a document titled Framework Agreement for Labour Intensive Construction which gives useful background information on the discussions that took place between important role players in the civil engineering industry, trade unions and civic organisations. Supporting initiatives are that of the Expanded Public Works Programmes (EPWP) which also support job creation by means of examining current work methods and revisiting these to create additional labour opportunities. It is not the intention of this or any other document to replace the traditional mechanised methods of construction, merely to enhance those construction practices by creating more labour employment opportunities. Further information can be obtained by accessing the following website: http://www.epwp.gov.za/Downloads/technical_legalguidelines.pdf

12.2.1 Objectives

- To maximise the use of LEC systems within public work programmes, with due regard to economics.
- To contribute to employment creation.
- To lead to the growth, stability and viability of the construction industries.
- To meet the social infrastructure needs of the people of South Africa.
- To ensure social infrastructure is equitably distributed and accessible.

The challenge facing the construction industry is how to create and adapt activities in such a way as to provide employment and entrepreneurial opportunities within the industry. To suit the multi-cultural societies of South Africa, it is necessary that we develop and adapt our own solutions to LEC and not just take methods directly from underdeveloped countries with different social structures.

LEC is not a new concept. In the early 1970s, the World Bank initiated a programme of work to explore the increased use of labour in construction, specifically road construction. LEC has been successfully implemented in Kenya, Nigeria, Zambia, Botswana, Lesotho and Mozambique.

It is important to obtain and to ensure continuous support from all the stakeholders in the project – from the national government through to the community where the project will be implemented. Without participation, the project could be a failure.

Participation by the community ensures ownership of the project by that community and will result in them taking pride in their achievements. It is important that, when seeking labour, the community leaders within a rural environment are approached first for their involvement and commitment. In many instances, this may result in the upliftment of the social conditions of that community. By involving the local community, one can establish what that community considers a priority in alleviating poor living conditions.

In South Africa, LEC has been applied particularly to the roads and transportation fields, because it is a known fact that roads form an integral part of our lives and provide a means of safe and economic transport of goods and people for various purposes. The provision of efficient and cost-effective transportation has a critical function to fulfil within the process of development in South Africa. An important aim of an optimum transportation policy should be to ensure the use of local labour resources in any civil engineering works so that the community gains maximum benefit from the construction activities and from the finished product.

While LEC is not the answer to all the country's problems, it does present an opportunity to alleviate unemployment, to increase the skills in the workforce, and to nurture emerging entrepreneurs.

Experience indicates that there are no technical reasons why LEC methods should not be successful.

12.2.2 Technical feasibility of LEC

Certain areas of construction are better suited to LEC than others – for example, street sweeping and litter collection. In urban and rural areas the cleaning of drainage channels is conducive to LEC and based on technical feasibility and cost effectiveness. However, feasibility is also related to the size of the project and the speed of construction. In Botswana and Kenya, labour-enhanced methods have been used successfully in the construction and maintenance of rural roads with low traffic volumes. A major drawback is the time necessary to carry out the initial technical studies, train staff and set up a competent organisation to cope with this new socio-economic and technical challenge.

12.2.3 National approach to labour enhancement in construction

Historically there have been two approaches to LEC. There has been a **short-term approach** (often in developed countries) where it has been used to stimulate employment levels. Such programmes are attempts to alleviate poverty, often in rural areas, where no alternative employment exists. The other approach is where labour has been employed (mainly casual labour) as a realistic and **economic alternative to machine-based methods.** In South Africa, some projects also aim to provide the 'locals' with necessary skills to become entrepreneurs.

The essence of LEC is to generate opportunities. The national need to create meaningful employment can be married with the need for social upliftment through LEC.

Future national prosperity in South Africa will be created, not inherited. It will not only grow out of the country's natural resources, but also out of our own labour pool. To increase our competitiveness, we need to build on skills, develop entrepreneurial opportunities and adopt a positive attitude to work. Government must also play its rightful role in the provision of the needs of the unemployed by creating a climate conducive to the creation of employment.

LEC is assessed on a number of factors, such as organisation, performance, suitability, productivity, equality and job design.

12.2.4 Organisation

LEC demands that organisation and labour become integral parts of the production process. Just as one would not run any construction activity without mechanics and workshops, one does not run a labour unit without organisation and training. LEC will require changes in project prioritisation, contract structure, tendering systems, community involvement, and construction. However, these changes can be accommodated within the time frame of normal project planning. It is also important that, when considering LEC contracts, the contracts be drawn up in such a way as to ensure that the expertise that has been developed in South Africa in construction contracting is not discarded. This means that the project management and technical skills of existing contractors must be utilised.

12.2.5 Performance

The performance of any engineering structures is critically dependent on the quality of construction. The industry has invested considerable effort in developing skills, plant and equipment to ensure quality construction.

12.2.6 Suitability of labour enhancement

The suitability of LEC for any activity depends, to a large extent, on quality issues. It is essential that the correct project is chosen for LEC and this will depend on the type of work that has to be undertaken, the size of the contract and the speed of the construction.

12.2.7 Productivity

It is generally accepted in all engineering circles that the labour costs as a percentage of the total costs for any contract are higher when using LEC. It is therefore important that the labour employed is productive. The following criteria should be met in order to achieve maximum productivity:

- Adequate **financial reward** must be set, preferably through an incentive system.
- The task to be performed must be of such a nature and carried out under such conditions that the workers can **take pride** in their work.
- The workers must be given **training** in the correct methods of doing the task.
- The **workers' needs** in terms of nutritious foods and adequate shelter must be met.
- **Management** must be good and the workers must feel it to be so.

12.2.8 Quality

A labour enhanced project must embody appropriate design and tender specifications to ensure the most cost-effective level of employment, without departing from acceptable **standards of construction.** It is therefore important that the quality standard be set before the contract is undertaken and that regular and proper supervision is undertaken to ensure that these standards are met.

12.2.9 Job design

In designing jobs for LEC, it is logical to break the tasks into the very simplest levels to facilitate the use of unskilled labour. Unfortunately, such jobs are often so narrow and meaningless that supervisors and higher levels of management are sucked into the control and coordination of details. This will cause frustration rather than participation and result in less time for constructive long-term improvement.

12.2.10 Job design in an LEC environment

The principles of job design are:

- **Optimum variety of tasks within the job.** While multi-skilling (the ability to be efficient in a number of tasks) is important, too much variety can be inefficient for training and production, and a source of frustration for the worker. However, too little variety can be conducive to boredom or fatigue. The optimum level is that which allows the operator to take a rest from a high level of attention or effort for a demanding activity by working at another related but less demanding task.
- **A meaningful pattern of tasks that gives each job the semblance of a single overall task.** The tasks should be such that, although involving different levels of attention, degrees of effort or kinds of skills, they are interdependent. Given such a pattern, the worker can more easily find a method of working suitable to his or her requirements and can more easily relate to his or her job and to those of others.
- **Optimum length of work cycle.** Too short a cycle means too much finishing and starting. Too long a cycle makes it difficult to build up a rhythm of work.
- **Some scope for setting standards of quantity and quality of production and suitable feedback.** Minimum standards generally have to be set by management to determine whether a worker is sufficiently trained, skilled or capable to hold a job. Workers are more likely to accept responsibility for higher standards if they have some freedom in setting them, and are more likely to learn from the job

if there is feedback. They can neither effectively set standards nor learn if there is insufficient feedback.

- **Inclusion in a job of some of the auxiliary and preparatory tasks.** A worker cannot and will not accept responsibility for matters outside his or her control. Insofar as the preceding criteria are met, the inclusion of such boundary tasks will extend the scope of the workers' responsibility and make for a higher level of quality in the job. The tasks should include some degree of care, skill, knowledge or effort that is worthy of respect in the community. The job should make a perceivable contribution to the utility of the product for the consumer.

12.2.11 LEC in rural and urban locations

LEC is technically applicable to both urban and rural areas. The closer one approaches the industrial centres, the more economic the considerations will be to justify the use of labour-enhanced methods.

Overlaying the cost issue is that of speed of construction. The speed of construction using LEC will often be slower than with plant-based methods. This places a constraint on its application for time-sensitive construction activities.

12.2.12 Wage rates and payment systems

The two aspects to consider are **the amount to be paid** and **the system of payment** on a daily task basis. The importance of settling on the right wage cannot be overstated, for both practical and philosophical reasons. At the practical level, too low a rate will cause disruptions, low productivity and difficulty in attracting labour. Too high a level will reduce the cost competitiveness of labour enhancement. The use of labour brokers, possibly under a tender system, will allow the community to have an input into wage rates for each project. All parties must agree to the system of payment before starting the project. It is always a good idea to approach or involve the local municipality when considering projects in their area as they will have a good indication of the labour rates applicable in the area.

12.2.13 Training

The provision of a well-planned and coordinated training programme for all levels is an essential requirement for successful LEC. Training is not confined to technical training – it also needs to encompass a small contractors' development programme, covering the observational aspects of people management and financial management, and business aspects such as tendering and marketing.

Questions that need to be asked include:

- Can the country afford to have so many people unemployed?
- What is the cost to the nation of an unemployed person?
- Can LEC be effectively incorporated into construction projects?
- What is the role of trade unions, and national and local authorities?
- What is the attitude of small contractors towards the implementation of LEC?

12.2.14 The pitfalls of LEC

- **Poor quality contracts.** Because LEC has become a fairly specialised application, it is essential that the contracts are drawn up and managed properly to ensure efficient and effective delivery.
- **Broken contractual agreements.** Sometimes LEC contracts are broken or abandoned before completion (i.e. workers start jobs but do not complete them).
- **False claims for funding to purchase materials.** One example of mismanagement is when the same invoice is used for more than one contract to claim for materials purchased.
- **Inexperience (of contractors)** in running contracts.

When utilising local labour on any project, it is still necessary to conform to the Basic Conditions of Employment as established by the Department of Labour, such as:

1. It is understood that most LEC labour contracts are for temporary labour for no longer than 24 months.
2. Working hours as follows:
    a. 40 hours per week
    b. 5 days a week
    c. 8 hours per day
    d. No more than 55 hours in any working week.
3. Meal breaks – workers are not allowed to work for more than five hours without a break. No entitlement for payment during a meal break is allowed.
4. Daily rest period – is measured from the time a worker ends work on one day until the time the same worker assumes work on the next day.
5. Work on Sundays and public holidays – is only allowed in cases of emergency or security work.
6. Sick leave – this only applies to workers who work four or more days a week who then become eligible to claim sick pay. A medical certificate must be produced which is signed by a professional medical practitioner.

7. Workmen's Compensation – when workers are injured during the course of normal daily work, they are entitled to claim compensation.
8. Maternity leave – workers are entitled to take up to four months unpaid maternity leave, but are not entitled to any payment or employment-related benefits during maternity leave.
9. Termination of service – an employer may terminate the service of a worker provided there are valid reasons and after following due termination procedure OR if the employment contract expires.
10. Certificate of service – on termination of service, the worker is entitled to a certificate containing the conditions under which the employment took place, including any training received.

It is common practice during large contracts where local labour is employed to also engage the services of a community liaison officer (CLO). It is always beneficial to the contractor if the CLO is from the local community and knows the 'political' environment within the community. The CLO is usually employed under the contract and is supervised by the main contractor. The main function of the CLO is to:

- assist the contractor with the selection and recruitment of local labour;
- represent the local community when sourcing local labour; and
- facilitate communication between the contractor, community and engineer.

The CLO's job description must include the following tasks and responsibilities:

1. Be available on site on a daily basis.
2. Consult with the contractor and engineer about the labour requirements relating to skills required and the number of workers needed.
3. Assist in identifying suitable temporary local labour and keeping a local labour register.
4. Assist in screening local labour in terms of the requirements.
5. Attend all meetings involving the contractor and community.
6. Inform local labourers regarding their conditions of employment.
7. Assist in maintaining good relations between the workers, contractor and community.
8. Assist the contractor in the training needs of the temporary labourers.
9. Attend disciplinary proceedings to ensure fair and reasonable practice.
10. Keep a daily record of work and community liaison activities.
11. Perform any other duties as agreed with the engineer or contractor as may be agreed or deemed fit.

12.2.15 Conclusion

To make LEC successful, it is important that there are changes in attitude at all levels and the empowering of unskilled workers can result in the sustainability of communities and their projects. An important criterion of empowerment is training, which should be done to fulfil the needs of the contract and its labour force.

We should also note that there are several organisational, labour and training issues that need to be addressed before LEC can be used successfully. It is clear that decision making in terms of LEC is no different to that which applies to conventional construction methods. The important considerations are still quality and performance.

It is important then, for the successful implementation of LEC, that the structure and management of projects should receive attention.

The form of empowerment that LEC can provide is what South Africa needs as a means of alleviating poverty. LEC can work in South Africa.

12.3 Summary

The purpose of this chapter was to:

- Explain the concept of labour-enhanced construction (LEC)
- Demonstrate the application of LEC in South African conditions
- Identify the objectives and principles behind LEC
- Explain the national approach to LEC and its processes.

Self-evaluation 12

1. Complete the sentences:
   a. Labour-intensive construction can contribute to _______ creation.
   b. Workers must be given _______ in the correct methods of doing a task.
   c. A labour-enhanced project must embody appropriate _______ and _______ specifications.
   d. _______ rates of payment will cause disruptions, low productivity and difficulty in attracting labour.
2. State whether the following are true or false:
   a. For labour-enhanced construction to be successful, it must have the support from all stakeholders.
   b. Labour-enhanced construction was initiated in South Africa.
   c. Two approaches are applicable to labour-enhanced construction.
   d. Work tasks must be broken down into their simplest levels to facilitate the use of unskilled workers. ➤

3. Answer the following short questions:
   a. How can one ensure maximum productivity?
   b. List the main principles of job design as applied in LEC.
   c. How can the implementation of LEC meet the social infrastructure needs of South Africa?
   d. Why is it important that all stakeholders participate in LEC projects?
   e. Why is it important to employ a CLO on a project which involves the use of local labour?
   f. Explain briefly the responsibilities of the CLO.
   g. Name the basic conditions of employment that workers are entitled to when employed as temporary workers.

Answers to self-evaluation 12

1. a. employment
   b. training
   c. design and tender
   d. low
2. a. true
   b. false
   c. true
   d. true
3. a. see ref 12.2.7, page 251
   b. see ref 12.2.10, page 252
   c. your opinion
   d. your opinion
   e. see ref 12.2.14 page 254
   f. see ref 12.2.14 page 254
   g. see ref 12.2.14 page 254

Index

This is a subject index arranged in letter-by-letter order. Figures and illustrations are indicated in *Italics*, and *see* and *see also* references guide the reader to the access terms used.

Printed in the USA
CPSIA information can be obtained
at www.ICGtesting.com
LVHW080713020224
770698LV00004B/148